Textbook of

SOIL SCIENCE

An Introduction

THE AUTHOR

Dr. Vishal Goyal works as an Assistant Professor (Stage III) in the College of Agriculture, Department of Soil Science, CCS HAU, Hisar (Haryana). He was born on 22nd November 1975. He has over 15 years of experience in teaching, research and extension. He is the life member of Indian Society of Soil Science. He has authored over a hundred research papers/ articles of International and national repute. Also he has authored one book for the UG students regarding vermicomposting and worms and one practical manual for the course on soil physics. His field of specialization is Soil Physics and Soil fertility. He is the recipient of Gold Medal award for his best PhD thesis during 2004-05. He has been awarded by the commendation prize for best Ph D thesis presentation by Indian Society of Soil Science, New Delhi.

Textbook of

SOIL SCIENCE

An Introduction

—: *Author* :—

Vishal Goyal

2025

Daya Publishing House®

A Division of

Astral International Pvt. Ltd.

New Delhi – 110 002

ISBN : 978-93-5919-978-8

CCS HAU/PUB# 24-123

Publisher's Note:

Every possible effort has been made to ensure that the information contained in this book is accurate at the time of going to press, and the publisher and author cannot accept responsibility for any errors or omissions, however caused. No responsibility for loss or damage occasioned to any person acting, or refraining from action, as a result of the material in this publication can be accepted by the editor, the publisher or the author. The Publisher is not associated with any product or vendor mentioned in the book. The contents of this work are intended to further general scientific research, understanding and discussion only. Readers should consult with a specialist where appropriate.

Every effort has been made to trace the owners of copyright material used in this book, if any. The author and the publisher will be grateful for any omission brought to their notice for acknowledgement in the future editions of the book.

Published by : **Daya Publishing House®**
A Division of
Astral International Pvt. Ltd.
– ISO 9001:2015 Certified Company –
4736/23, Ansari Road, Darya Ganj
New Delhi-110 002
Ph. 011-43549197, 8130496929
E-mail: info@astralint.com
Website: www.astralint.com

Printed at : **Replika Press Pvt. Ltd**

Message

Soils are dynamic and diverse natural systems that lie at the interface between earth, air, water, and life. They are critical ecosystem service providers for the sustenance of humanity. The improved conservation and management of soils is among the great challenges and opportunities which need to be focussed for its sustainability. They store and provide water for plants. They prevent floods by transferring water slowly to streams and groundwater. They filter and remediate pollutants. They cycle and recycle nutrients and wastes — transforming them into biologically available forms, storing them away for later use, and preventing their leaching to ground and surface waters. Soils provide habitat for a vast diversity of life. They take up and release important gases, including oxygen and greenhouse gases. Many of these ecosystem services are being lost through the degradation and loss of soils. The conservation, restoration, and optimization of ecosystem services provided by soils is among the great challenges for humanity in the 21st century.

The education in soil science and knowledge of soil is important to increase the students' participation. Soil Science, as a general feeling 'a tedious subject', needs orientation of the subject matter in a simple language and lucid manner to make it perfect learning aid and to create an interest among the students. This introductory book in soil science introduces the student to the subject management, and conservation of soils as natural bodies, as media for plant growth, and as components of the larger ecosystem. This book presents basic concepts of all aspects of soil science including; composition and genesis; physical, chemical, and biological properties; soil water, air and temperature; classification and mapping; soil reactions; management practices; and soil fertility and productivity. It introduces the relationships of soil to current concerns such as environmental quality. This book should create awareness of soil as a basic natural resource, the use or abuse of which has a considerable influence on human society and life in general.

The book has been well written covering all the basic requirements laid down in the syllabus for undergraduate and post graduate students of agriculture. There is a great need for such book that covered comprehensive information on all basic facts and aspects of soil science. The information is presented in most readable form with figures and diagrams wherever necessary. With a focus on practical applications, the book provides readers with a solid foundation in the principles of soil science and equips them with the knowledge and skills needed to understand and manage soil resources effectively.

I congratulate the author for his efforts and wish him all the best.

Prof. B.R. Kamboj

Vice Chancellor
CCS Haryana Agricultural University
Hisar (Haryana)

Foreword

Soil Science is the study of soil as a natural resource on the surface of the Earth including soil formation, classification and mapping; physical, chemical, biological, and fertility properties of soils; and these properties in relation to the use and management of soils. Throughout human history, our relationship with the soil has affected our ability to cultivate crops and influenced the success of civilizations. This relationship between humans, the earth, and food sources affirms soil as the foundation of agriculture. It is clear that agriculture sustains and defines our modern lives, but it is often disruptive of natural ecosystems. This is especially true for plant communities, animal populations, soil systems, and water resources. Understanding, evaluating, and balancing detrimental and beneficial agricultural disturbances of soil and water resources are essential tasks in human efforts to sustain and improve human well-being. Such knowledge influences our emerging ethics of sustainability and responsibility to human populations and ecosystems of the future.

Soil is frequently referred to as the "fertile substrate", however, not all soils are suitable for growing crops. Ideal soils for agriculture are balanced in contributions from mineral components (sand: 0.05–2 mm, silt: 0.002–0.05 mm, clay: <0.002 mm), soil organic matter (SOM), air, and water. The balanced contributions of these components allow for water retention and drainage, oxygen in the root zone, nutrients to facilitate crop growth; and they provide physical support for plants. The distribution of these soil components in a particular soil is influenced by the five factors of soil formation: parent material, time, climate, organisms, and topography. Each one of these factors plays a direct and overlapping role in influencing the suitability of a soil for agriculture.

Students of the UG and PG in agricultural subject must be aware of the basic knowledge of the processes taking place in the soil which can be obtained from the basic text book of soil science. This book aims to provide the fundamental understanding of the nature and properties of soils, an ability to apply knowledge in academic and real world contexts, such as food security or environmental sustainability, and opportunities to learn from world renowned researchers from a range of specific topics in basic soil science.

Dean

College of Agriculture
CCS Haryana Agricultural University,
Hisar (Haryana)

Preface

The sustainable agricultural production is pre-requisite to ensure the food security of growing population of the country. The agricultural production primarily depends upon the soil quality and productivity. The maintenance of soil productivity requires proper understanding of the soil properties and processes taking place in soil and their effects on plant growth. This text book describes the different factors and process of soil formation along with the physical chemical and biological properties of the soil to assess their quality for potential use for optimum plant growth.

Moreover, Soil Science is a favorable subject for UG and PG students of agricultural stream. This book will be useful for undergraduate and postgraduate students of Agricultural Universities and will cover almost all the aspects of soil science. An effort has been made to cover various aspects of Soil Science and its basics such as weathering of rocks and minerals for the formation of soil, soil morphology, and soil physics, chemistry, fertility and soil pedology viz. soil water, air and temperature, soil classification, soils of India, Soil colloids, soil reaction etc. It would also be helpful for students for the preparation of various competitive examinations like JRF, SRF, NET, ARS, AO (Agriculture Officer), UG and PG entrance.

The books and the literature we consulted are mentioned along the text and the source is also mentioned alongwith. We are highly thankful to our university's Worthy Vice-chancellor, Prof. B. R. Kamboj, for providing a platform to create the manuscript. Last but not least, we sincerely thank Dr. S.K. Pahuja, Dean, COA, Prof and Head, Dr. Manoj Kumar Sharma, Department of Soil Science and Dr. O.P. Chaudhary, Principal, COA, Kaul, CCS Haryana

Agricultural University, Hisar for their valuable suggestions and their kind support during the preparation of this book. The interest taken by the publisher in this book is also highly appreciated.

Dr. Vishal Goyal

Contents

Chapter 1

Soil and the Land

Land: Land may be defined as that part of the earth which is not covered by water resources permanently. It is a two-dimensional body that constitutes geographical area and landscape.

Soil: Soil may be defined as the three-dimensional entity with length, breadth and depth and is hidden below land surface

Land as a resource is the dry part of the earth that is not under water cover permanently. Soil, on the other hand, is the a complex mixture of different materials such as organic matter, microorganisms, minerals, water and air that forms on the surface of land and supports plant growth and life.

The land however may or may not support life, but the soil supports life through vegetation.

WHAT IS SOIL?

Soil is a 3-D, natural, dynamic body constituting mineral as well as organic matter occurring on the surface of the earth, differentiated into horizons of varied depth, developed by processes of pedogenesis resulted during weathering of rocks differing from parent material lying below, possessing definite physico-chemical and biological properties which are the result of the effect of climate and living organism on parent material modified by relief over a period of time.

Definition of soil varies according to its end user-

Pedologist: According to pedologist, soil as a natural dynamic body which is differentiated into horizons formed by pedogenic processes which supports life on it.

Engineer: According to engineer, soil is an unconsolidated material overlying the bed rock.

Geologist: According to geologist, soil is a natural medium for the growth of plants on land.

Microbiologist: According to microbiologist, soil is a medium which supports life on the land

The reason for studying soils by different peoples is to get an idea of what the soils are and how they can be made to be used for a particular purpose so that it can cater to the need of the people i.e. engineers, farmers, geologist, home owner etc. It is also a requisite for determining how best to conserve soils for future generation.

FUNCTIONS OF SOILS

1. It support plant growth
2. Controls the fate of water in the hydrologic system
3. Functions as a nature's recycling system
4. Habitats for living organisms
5. An engineering medium

APPROACHES OF SOIL STUDY

1. **Pedological approach:** Pedology has been derived from Greek word *pedon* which means soil or earth and logy means its study. So, pedology is the study of soil in its natural environment dealing with the genesis (formation), morphology and classification. This includes study of rocks and rock forming minerals, factors and processes of soil formation. So, pedology is important for planning of towns, engineering purposes and to the farmers and agricultural Scientists.
2. **Edaphological approach** (Greek word *Edaphos* means ground**):** The science that deals with the influence of soils on living things particularly plants including man's use of land for plant growth. This is the applied branch of soil science which involves study of properties of soil which affects plant growth, nutrition and yield of the crop. This approach deals with soil fertility, chemistry, physics, microbiology etc.

   So far, out of 115 elements of the periodic table, there are 17 essential nutrients needed by the plants to survive and complete its life cycle. These nutrients are

   - Carbon (C), Hydrogen (H) and Oxygen (O)obtained from Air
   - Nitrogen (N), Phosphorous(P) and Potassium (K)are the macro nutrients (primary nutrients) required by the plants in larger quantities
   - Calcium (Ca), Magnesium (Mg) and Sulphur (S) considers as the secondary nutrients because of their moderate requirement by the plants.

- Zinc (Zn), Molybednum (Mo), Copper (Cu), Iron (Fe), Manganese (Mn), Chlorine (Cl^-), Cobalt (Co,) and Nickel (Ni)are the micronutrients which are required in very minute quantities by the plants but without which plant cannot be able to complete its life cycle.

Soil as a Natural and Dynamic body

Soil is a natural dynamic body which is formed under various ecosystems with the help of factors and processes which reacts on the rocks and minerals to form the soil. Various factors and processes combinedly influence the formation of soil which is continuous process and is still going on.

Soil as a Porous System

Soil is considered as a porous system as it is made by combining together of the solid particles in to its different size aggregates which lead to the formation of pores (macro as well as micro pores). These pore spaces may constitute liquid or gases in varying amount depending upon the environmental condition and climate of the area. This pore spaces permits the movement of nutrients through transportation of water and air through it.

Soil as a living body

Soil is called the living body since it contains micro and macro-organism in varying proportions which in-turn helps in conversion of organic matter into its stable form which is called humus. Most of the soils except that of forest and swampy soils contain 1-2 % of organic matter and are made up of in organic alumino-silicate minerals, still it supports life in the soil.

A FIELD VIEW OF THE SOIL

Surface soil and subsurface soil

Surface soils are considered as those soils from which most of the field crops take nutrients for their growth and proliferation. When the soil is plowed and cultivated the upper 0-15 cm is tilled. This soil is called as surface or top soil which may also be called as furrow slice. Surface soil is the root development zone for most of the field crops which may cater the need of nutrients and water to the plants for their growth. Surface soils are more fertile than the subsurface soils due to incorporation of organic residues and commercial fertilizers, thus allowing the soil's fertility and productivity.

Thus, *fertility* refers to the inherent capacity of soil to supply suitable quantities of mineral nutrients from different sources viz. nitrogen, phosphorous, potassium, calcium, magnesium sulphur and certain micronutrients needed by the plants for their growth. *Productivity* refers to the ability of the soil to produce the crop. A fertile soil cannot always be productivity if it has a poor structure but a productive soil is always fertile.

The subsoil on the other hand comprised of those soil layers underneath the top soil. This soil layer is more compacted as it cannot be disturbed by tillage. Plant root activity is very less and nutrient content of the subsoil is also less although the moisture content of the sub soil may be high. The subsoil due to its different structure decides the movement of water and nutrient in the surface soil. Downward movement of drainage water is sometimes impeded due to impervious structure of the sub soil. The land-use should be properly decided based on nature of subsoil than on the top soil characteristics.

Soil and Regolith

Regolith is the unconsolidated material overlying the rocks which may be shallow in depth or of several meter thick. The material has been weathered from the underlying rocks and may be formed by transportation of material from the other places due to water, wind, and/or air and consolidate to form regolith. Thus, regolith varies from place to pace in its composition. The upper few centimeter of the regolith differs from the regolith below. It is subjected to high intense weathering than the lower portion. This intense weathering may give rise to horizons of the regolith which varies in the characteristics from the layer overlying it. It is also higher in organic matter content and has higher microbial activities. The so formed upper and bio-chemically weathered part of the regolith is called as ***soil or solum***.

Composition of Earth's crust

- Each layer has a distinctive composition i.e. physical as well as chemical, crust being quite different from the Earth as a whole. In whole Earth, Fe, O, Si and Mg constitute about 93% in composition while crust composition constitute Si, O & Al in about 82%.
- The earth's crust is composed of mineral matter consisting of elements. All elements known (117, only 92 are naturally occurring) are found in earth's crust in various combinations to form compounds/minerals (about 4000). The earth's crust is composed of almost all the mineral matter but only a few minerals occur in abundance. Oxygen is most abundant element (46%) which usually occurs in combination with other elements; followed by silicon, Si (about 28%); aluminium, Al (about 8%); iron, Fe (about 6%); magnesium, Mg (about 4%); calcium, Ca (about 2.4%); potassium, K (about 2.3%); sodium, Na (about 2.1%) and remaining elements constitute about 1%.

Chemical composition of Earth

(Source: https://www.shutterstock.com/search/earth-crust)

- Out of known elements, above mentioned 8 elements are sufficiently abundant constituting 98.8% by weight of the earth's crust. Oxygen (O) and silicon (Si) (non-metallic) comprise nearly three-fourth of the total crust composition. The other 6 elements are metals.

Chapter 2

Rocks and Minerals

The earth was formed due to cooling and subsequently crystallization or consolidation of melted rocks (molten magma) occurred on the earth. During cooling, elements present in molten mass recombine naturally in an orderly way and consolidated into huge masses arising varieties of rocks and minerals.

Rocks are the masses of hard substance composed of either one or more than one mineral which has been brought together cohesively during rock-forming processes and forms the essential part of earth's crust i.e. formed from the solidification of molten magma (melted rocks).

CLASSIFICATION OF ROCKS BASED ON COMPOSITION

1. Mono-mineralic rocks

"The rocks which are dominated by single mineral only are called as *mono-mineralic rocks*". Examples: quartzite, dolomite, limestone rocks, etc. Quartzite is the rock which contains only quartz mineral. Dolomite and limestone contains calcium carbonate minerals.

2. Multi or poly-mineralic rocks

"The rocks which are dominated by two or more than two minerals in their composition are classified as *multi-mineralic or poly–mineralic rocks*". Examples: granite, basalt, gabbro, sand stone, etc. **Granite** is a multi-mineralic rock consisting of quartz, feldspar and mica, **Sandstone** is the rock consisting of quartz, feldspar along with silt and clay, **Basalt** and **Gabbro** mainly constitutes feldspar, pyroxenes with little amount of olivines.

Minerals are the naturally occurring inorganic crystalline solid having an orderly internal arrangement of elements or atoms (i.e. having characteristic geometric form), possessing definite physical properties and chemical composition.(*A crystal* or *crystalline solid* is a *solid* material whose constituents (such as *atoms, molecules, or ions) are arranged in a highly orderly microscopic* *structure, forming a crystal lattice* that extends in all directions and arranged with characteristics angles between them).

The minerals may be composed of one or more than one elements but most minerals consist of two or more elements. Examples: quartz (SiO_2), orthoclase ($KAlSi_3O_8$), olivine [$(Mg, Fe)_2SiO_4$], calcite ($CaCO_3$), gypsum ($CaSO_4.2H_2O$), etc. Minerals consisting of only one of the naturally occurring elements may be a metal, such as copper, iron, aluminium or a non-metal such as silicon, carbon, sulphur.

Depending on their mineralogical composition, the rocks consist of certain characteristics such as structure, colour, specific gravity and mineralogical composition etc. for their identification.

Petrology: The science that deals with the study of rocks that forms the earth's crust is termed as *petrology*.

It is classified as *petrography* and *petrogenesis*.

Petrography is the study that accords with the description of rocks.

Petrogenesis (Petro- rocks and genesis- formation) is study of formation or genesis of rocks.

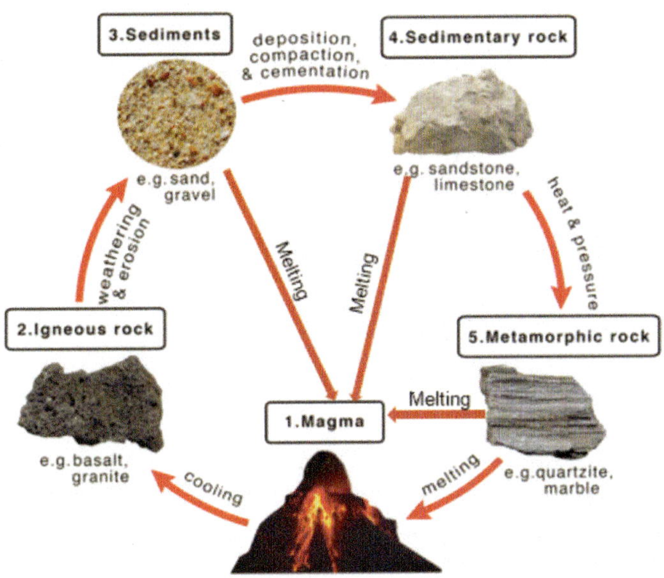

Rock Cycle

(Source:https://www.sciencefacts.net/wp-content/uploads/2020/04/Steps-of-the-Rock-Cycle.jpg)

Classification of rocks based on mode of formation

On the basis of mode of formation, rocks are classified in to three categories viz.

i) Igneous (95%)

ii) Sedimentary (4.9%)

iii) Metamorphic rocks (0.1%)

Igeneous rocks constitute the largest part of the rocks covered on the earth i.e. 95% of the weight of the lithosphere while sedimentary and metamorphic rocks constitute only 4.9 and 0.1 % of the weight of the lithosphere.

I. Igneous rocks (*Ignis* means fire)

Those rocks which are formed by cooling and crystallization of molten magma on or beneath the earth's surface are called as igneous (primary) rocks. (*Magma is a complex hot mixture of silicates containing water vapours and gases whose temperature range from 700-1400^{0}C and originating from deep in the earth*).Molten magma flowing on the surface of earth is also called as lava.

Characteristics of Igneous rocks

i) Igneous rocks are non-laminar (do not arranged in layers) having massive or compacted structure.

ii) They are the source of parent material for other rocks and for soils too.

iii) They mainly constitutes primary minerals like quartz, feldspar, mica, pyroxenes and thus also known as primary rocks

iv) They do not include any fossil deposits.

v) They may vary in size and forms which depends on the rate of cooling of the magma. When the magma cooled down and solidified at the surface of the earth very rapidly, there is less time for the formation of crystals, thus the rocks formed are fine-grained and lacks crystalline structure. Eg. Obsedian, Basalt. If magma cools and consolidate (solidify) at a very slow rate inside the earth, there is enough time for the crystallization, and thus igneous rocks formed are characterized by coarse grained crystalline structure Eg. Granite.

TYPES OF IGNEOUS ROCKS

1. Volcanic or extrusive rocks

Those rocks which are formed due to cooling and consolidation of magma on the surface of earth are called as volcanic or extrusive igneous rocks. Examples: basalt, rhyolite, andesite, obsidian, pumice etc.

As the magma comes to the surface, the pressure is released and the water turns into steam. Sometimes, the steam cannot escape and it forms a globular/ spherical cavity within the rocks. The globular cavity is known as *vesicle*. The rocks which contain vesicles are known as *vesicular rocks*. The vesicles are often filled with some minerals such as quartz, calcite, etc. and are known as *amygdaloidal rocks*.

The volcanic rocks may be highly porous due to escaping of volatile constituents. Example: ***pumice** and **scoria.*** The pumice is very light weight rock due to porous structure. Its specific gravity is less than unity which makes it to float on water. However in composition, it is comparable to granite/rhyolite.

Scoria

Pumice

(Source: https://www.britannica.com/science/pumice)

Basalt constitute more than 90 % of all the volcanic rocks found on earth as it has low silica content and low viscosity, thus there is quick flow of lava that spreads to a greater area before cooling and solidification.

2. Plutonic or intrusive rocks

When the pressure subsides within the earth's crust, the magma starts moving back and on its way cools down to form rock beneath the earth's surface. Such rocks are called as plutonic or intrusive rocks. Examples: granite, gabbro, diorite, etc.

Among the intrusive rocks, the rocks which are consolidated at a much greater depth are called *plutonic rocks*. Examples: granite, diorite, etc.

Sometimes, the molten mass is forced into the cracks of the pre-existing rocks and a new rock is formed. The rock consolidated in vertical cracks forming wall like masses are called *dykes*. The rocks formed in horizontal cracks are known as *sills*where as the rocks formed in irregular and narrow cracks are called *vein*.

Composition of igneous rocks: Igneous rocks are made up of mainly primary minerals (92%) consisting of feldspars (58%), amphiboles & pyroxenes (16%), quartz (13%), micas (4%), etc

Classification of igneous rock: Based on the silica content, the igneous rocks are classified as:

a) Acidic : 65-75%, - granite, pegmatite, rhyolite, etc.

b) Neutral : : 55-65%, - diorite, syenite, andesite, etc.

c) *Basic* : 45-55%, - basalt, gabbro, dolerite, etc.

d) *Ultra-basic* : <45%, - olivine, peridolite, etc.

Important Igneous rocks are:

a) *Basalt:* It is most abundantly formed fine grained dark colored basic rock mainly composed of feldspars (50 %) and ferromagnesian (50 %) minerals. Gabbro has same composition but is coarse grained rock.

b) *Granite:* A coarse grained textured acidic rock and light coloured due to large amount of Feldspar (ranging from 60-70 %), quartz (ranging from 20-30 %) and ferro-magnesian minerals (constituting 3-10%).

c) *Gabbro*: A coarse grained textured rocks formed beneath the earth's surface and constitutes essentially ferro-magnesian minerals.

Basalt

Granite

Gabbro

II. Sedimentary rocks (sediments – settling)

These rocks are formed from the weathered products of igneous rocks. Different agencies water, wind, glaciers and/or gravity carry away the sediments to other place where it get deposited and cemented or consolidated to form sedimentary/ secondary rocks.

The most striking feature of the sedimentary rocks is the stratification i.e. the deposition of sediments in layers one over the other and the rocks so formed are called as ***stratified rocks***. The formation of sedimentary rocks takes place through four stages:

i) Weathering,

ii) Transportation

iii) Deposition or sedimentation and

iv) Cementation and compaction (known as diagenesis)

(*Diagenesis refers to the transformation/conversion of unconsolidated sediments to hard rock through compaction and cementation*).

The common examples of sedimentary rocks are:

a) ***Sandstone*:** Mainly composed of quartz with cementing agents as $CaCO_3$, iron oxides and clays and feldspar due to which they are resistant to weathering. It may vary in colour from light to red coloured due to impurities, usually granular and porous structure.

b) ***Shale*:** Mainly composed of clay minerals, quartz and cementing agents ($CaCO_3$, iron oxides and clays). It has thinly laminated fined grained structure and light to dark coloured. Shale has the tendency to split into thin layers (laminae) of approximate one centimeter or less.

c) ***Limestone*:** Composed of calcite mineral with little iron oxides, clays, phosphate and organic matter. Usually of light grey to yellow having fine grained and compact structure. It is a soft rock which can be easily scratched and effervescence readily with acid.

d) ***Conglomerate*:** Fine grained sediments, highly variable in mineral composition, shape, structure, colour and texture. Mainly comprised of quartz, feldspars along with the sediments of varying sizes.

Sandstone

Shale

Limestone

Conglomerate

(Source: https://www.sandatlas.org/)

Sedimentary rocks constitute 4.9% of lithosphere (shale 4.0%, sandstone 0.7%, limestone 0.2%).

Classification of sedimentary rocks on the basis of mode of formation

The sedimentary rocks are classified into six types.

1. ***Arenaceous rocks:*** The rocks formed due to deposition of coarse grained particles carried by the water are called as arenaceous rocks. Depending upon the nature of cementing materials, some rocks are hard but most of the rocks are loose and undergo weathering easily. Examples: *sandstone, conglomerate,* etc.
2. ***Argillaceous rocks:*** These rocks are formed from small sized particles composed of hydrated silicate of alumina (clays), calcareous materials, and other silicates. The deposition of clay mainly consisting of silicate of alumina is known as *Kaolin or china clay.* Examples: *mudstone, shale,* etc.
3. ***Calcareous rocks:*** Calcareous rocks consist of carbonate of calcium and/or magnesium.These rocks may be formed by deposition and compaction/cementation of sediments, chemical precipitation or organic agency. These rocks are mainly composed of debris from plants and animals when formed by the organic agency. When rocks are formed by chemical precipitation, the calcareous materials get deposited in form of sheets or layers. Initially, the precipitate is soft but soon after it becomes hard. Examples: *limestone, dolomite,* etc.
4. ***Carbonaceous rock:*** These rocks are formed by decomposition of vegetation in an anaerobic condition. When plants decompose under anaerobic conditions as under water or in lower layers of earth, major proportion of its carbonaceous matter is retained and slowly converted into coal. Depending on the extent of change, different carbonaceous rocks are formed, e.g., *peat, lignite, coal, anthracite,* etc.
5. ***Siliceous rocks:*** These rocks have microscopic silica as the principal constituent. These are dense, hard, and fine-grained. They are commonly formed from silica-secreting organisms such as radiolarians, diatoms, or some types of sponges. Example: *Chert*
6. ***Precipitated rocks:*** These rocks are formed from deposits formed by cooling, evaporation or by direct chemical precipitation. These rocks form deposits ranging from few inches to several feet. These rocks are found underground or on the surface. Examples: *Oxides- haematite, limonite, bauxite, quartz; carbonates- limestone; sulphates- gypsum; phosphates-phosphrite; and chlorides- rock salt.*

III. Metamorphic rocks (metamorphism- change in form)

The metamorphic rocks are formed when igneous and sedimentary rocks undergo a physical and/or chemical change by the action of internal heat and/or pressure. These changes alter the physical and chemical equilibrium of the mineral assemblage/grouping, resulting in metamorphic changes and a new equilibrium is reached. The constituent minerals of the rock are changed into more stable minerals that have better structural arrangement. The structure and mineral composition of metamorphic rocks depend on the composition of the original rocks and the kind of metamorphism.

The metamorphic rocks formed by the action of heat are known as *thermal metamorphic rocks* whereas the rocks brought out by the action of pressure are known as *dynamo metamorphic rocks* while the rocks formed by the combined action of heat and pressure are called as *dynamo-thermal metamorphic rocks. (Serpentinite is a metamorphic rock having serpentine group of minerals)*

Classification of metamorphic rocks: On the basis of structure and composition, metamorphic rocks can be grouped as *foliated* and *non-foliated* metamorphic rocks.

i) ***Foliated metamorphic rocks:*** These rocks are characterized by foliation (*arranged in parallel or sub-parallel pattern, thick in the middle and thin at the edges*). Bands of alternative dark and light coloured minerals are characteristics of the foliated metamorphic rocks. Examples: gneiss and slate.

ii) ***Non-foliated metamorphic rocks:*** The non-foliated metamorphic rocks consist of minerals which are more or less equi-dimensional in their shape. Examples: marble and quartzite.

Some important metamorphic rocks:

1. ***Gneiss:*** Igneous rock such as granite is commonly modified to gneiss due to the action of heat (thermal metamorphism) in which light and dark minerals are reoriented into bands. Gneiss has foliated structure.
2. ***Marble:*** Sedimentary rocks such as limestone or dolomite are metamorphosed to marble due to action of heat. It has compact and non-foliated structure, and found in various colours i.e. white, red, green, black, etc.
3. ***Slate:*** Shale- a sedimentary rock is metamorphosed to slate due to action of pressure. The colour of slate is grey to black having foliated structure.
4. ***Schists:*** Schist rocks have foliated structure andare formed from basalt and shale due to action of pressure. Schists are light to dark in colour having foliated structure.

5. ***Quartzite:*** Quartzite is formed from the sandstone due to action of heat and pressure. The quartzite is extremely resistance to weathering and produces shallow light textured sandy soils. The quartzite has non-foliated structure.

Gneiss

Marble

Slate

(Source: https://www.etsy.com/in-en/listing/747157822/black-gray-slate-metamorphic-rock-3)

Schist

Quartzite

(Source: https://geologyscience.com/rocks/metamorphic-rocks)

Some of the metamorphic rocks with their characteristics are:

Rock	Composition	Structure	Colour	Metamorphism
Gneiss	Granite (igneous rock)	Foliated	light and dark bands	Heat
Marbles	Lime stone/ dolomite (sedimentary rock)	Non-foliated	white, red, green, black.	Heat
Slate	Shale (sedimentary rock)	Foliated	grey to black	Pressure
Schist	Basalt + Shales (igneous + sedimentary rock)	Foliated	light to dark	Pressure
Quartzite	Sand stone (sedimentary)	Non-foliated	light to red	Heat and pressure

SOIL FORMING MINERALS

General characteristics of minerals:

1. ***Colour***: Mineral colour is determined by absorption and reflectance of light through it. Although, colour is not always a distinguishing characteristic for identifying minerals. Impurities may also lead to variations in a mineral's color. Mineral may be white, black, pink, yellow or green and even colourless. Examples: quartz-colourless, calcite-white, magnetite-black, orthoclase-pink, iron pyrite-yellow. Few minerals like quartz, calcite, halite, and fluorite occurs in a variety of colors, and other minerals, such as olivine, malachite, and amphibole occurs in a limited colors and some minerals may have the same colour too.

Quartz

Calcite

Halite

(Source: htttp://Soils4teachers.org/mineralogy)

2. ***Luster***: It is the shining of an individual mineral. Luster is how the surface of a mineral reflects light. Luster is mainly of two types: metallic and nonmetallic. Opaque and shiny minerals have a metallic luster eg. Pyrite. Few minerals also have a non-metallic luster eg. Quartz.

Non-metallic luster is of various types:

i) Vitreous : like ordinary glass, e.g., rock salt/crystal

ii) Pearly : like pearl, e.g., zeolite

iii) Earthy : dull appearance resemblance to earth (gypsum)

3. ***Light transmission***: The mineral having the ability to transmit light through it is the light transmission ability. Minerals through which light can pass easily are called as *transparent mineral* e.g., quartz. *Opaque minerals* are those through which light cannot pass e.g. magnetite, iron pyrite, copper pyrite etc. However, minerals that allow a only limited light to pass through them are known as *translucent mineral* e.g., milk quartz.

4. ***Hardness***: Relative hardness of the minerals can be measured on Mohs' hardness scale. It is the resistance of a mineral to scratching. German geologist ***Frierich Mohs*** (1812) has developed a scale known as Mohs's scale of hardness for different minerals. Relative hardness

(1 to 10) of some minerals in order is: Talc (1 i.e. softest), Gypsum (2), Calcite (3), Fluorite (4), Apatite (5), Feldspars (6), Quartz (7), Topaz ($Al_2SiO_4(F,OH)_2$) (8), Corundum (Al_2O_3)(9) and Diamond (10 i.e. hardest).

(Roughly, the hardness of a finger nail 2, glass piece 5.5, and sharp steel (knife) is 6.5)

5. ***Streak:*** It is colour of line drawn by rubbing the mineral against a coarse hard plane surface. Streak is the moist reliable property of identifying minerals than colours since streak does not vary with minerals. Generally, white minerals have white coloured streak. Streak of some of the important minerals. For eg. Rock salt mineral is white in colour will produce white streak on rubbing. Orthoclase mineral pink in colour produces white streak on rubbing while Iron pyrite yellow in colour produces green colour streak, magnetite black colour will produce black colour streak.
6. ***Specific gravity:*** Specific gravity refers to the mass of a mineral. Specific gravity is the qualitative means of expressing the characteristic of minerals and is expressed as the ratio of the density of mineral w.r.t. the density of reference substance (Water). The minerals differ widely in their specific gravities. Some minerals are very light having specific gravity of 2.65 (quartz, feldspar, mica etc.) while some may have specific gravity even >5.0 (haematite, magnetite, pyrite etc.).
7. ***Forms and structure:*** A mineral having a definite arrangement of its constituent elements is called *crystalline mineral*. Examples: olivine, orthoclase, quartz, calcite, etc. When mineral does not have a definite arrangement of its constituents is called *amorphous*. Examples: haematite, gibbsite.

Further, depending on the size of axes and angles between the axes of the crystal, a mineral may be subdivided into six different forms:

i) ***Monometric (cubic or isometric):*** All the three axes of a mineral are of similar size (length and breadth) and each of angle are at 90° to each other, e.g., rock salt (NaCl). i.e. $A = B \neq C$ All Angles are of 90°	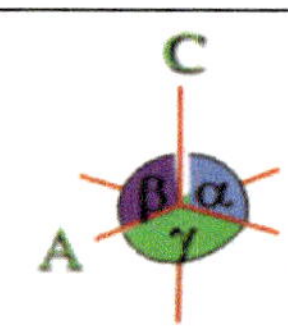 (***Axes and angles***) 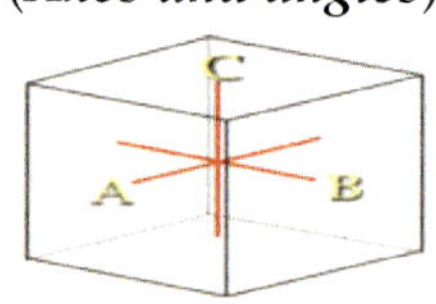

ii) ***Dimetric (octahedron or tetragonal)***: Two of the axes are equal but all the three angles are right angle, e.g., zircon (ZrO_2. SiO_2). A= B1 C All angle are of 90°	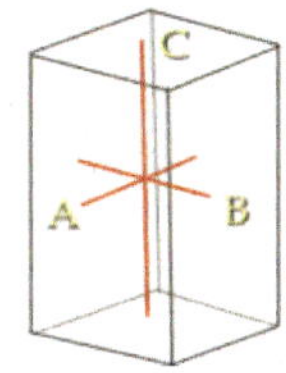
iii) ***Trimetric (orthorhombic):*** When all the axes are not equal to each other but the angles are at right angles to each other. e.g., olivine [$(Mg, Fe)_2.SiO_4$]. A^1 B^1 C Angles are of 90°	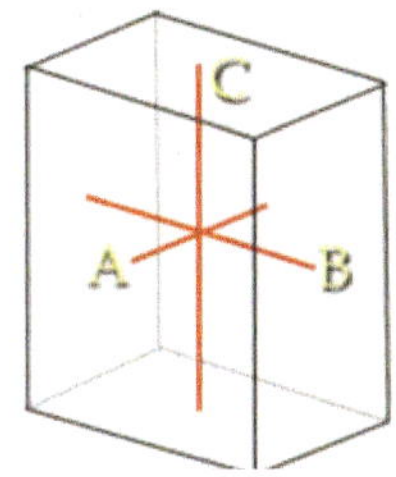
iv) ***Monoclinic***: When all the three axes are unequal, and two of which are at right angle to one another while the third axis makes right angle to the plane of the other two axes, e.g., orthoclase, gypsum, hornblende, etc. Axes Length Relationships: A B C Angle α Angle γ = Angle β = 90°	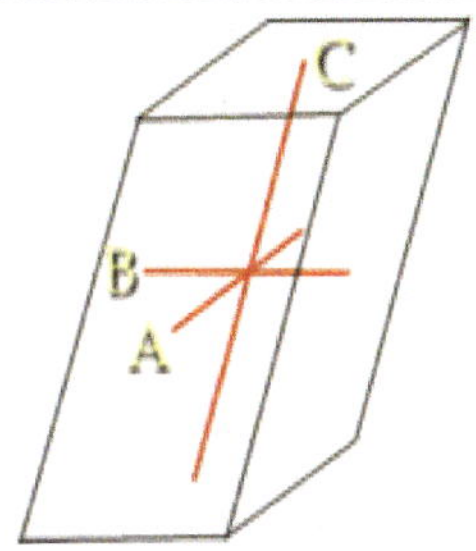
v) ***Triclinic***: All the three axes are unequal and none of them forms a right angle with one another. Example: The minerals of plagioclase group of feldspars such as Na-feldspar (albite) i.e. [Na_2O $Al_2O_3.6SiO_2$]. A B C Angles α Angle β Angle γ	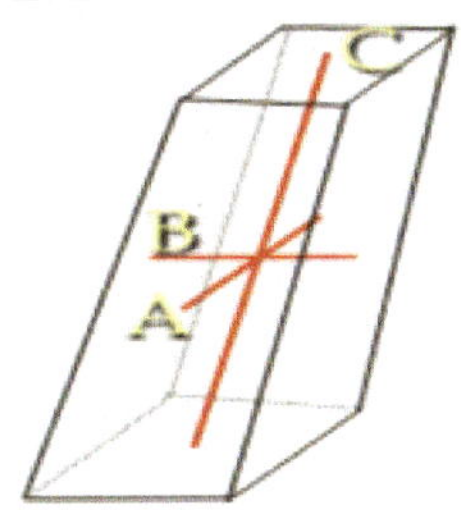
vi) ***Hexagonal***: There are three equal axes at 120° and one more fourth axis of a different length at right angle to the other three, e.g. quartz, calcite. Axes length relationships: A = B = C D Angles α = Angle β = 90° and Angle γ = 120°	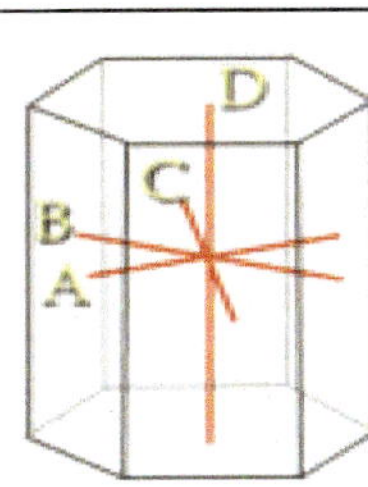

The arrangement of crystals of mineral gives rise to several kinds of pattern called structure such as: granular, laminar (plate like) and columnar (needle or pillar like)

8. ***Cleavage:*** Crystals often contain the planes of atoms along which the bonding between the atoms is weaker than along other planes. In such a case, if the mineral is struck with a hard object, it will tend to break

along these planes. This property of breaking along specific planes is termed ***cleavage.*** So, cleavage is the tendency of a mineral to break along planes of weakness in the chemical bonds sometimes parallel to crystal faces and yielding more or less smooth surfaces. If mineral break along the irregular surface, it does not have cleavage i.e. plane of weakness.

Eg. *Quartz* consists of extensively interconnected silica tetrahedral and does not have crystallographic weakness whereas mica consists of sheets of silica tetrahedral which are weakly held together.

9. ***Fracture:*** Fracture refers to the non-planar breakage of minerals i.e. discontinuities in displacement along the surface. The fracture may be conchoidal, even or uneven. A fracture is *conchoidal* if the resulting surface is more or less curved. If the surface of the fractured plane is smooth and plane, they are called *even* otherwise *uneven.*
10. ***Tenacity:*** It is the resistance of the mineral to break, bends and crushed against an external force on it. A mineral may be brittle (breaking into small pieces), flexible (easily bends and does not come to its original shape), and malleable (flattens upon applying external force).

Classification of soil forming minerals

The minerals may be classified on the basis of their mode of formation/ origin, abundance (amount of mineral in the rock), specific gravity and chemical composition.

1. ***Classification of minerals based on mode of formation/origin:*** The minerals are categorized into two groups on the basis of their mode of formation.
   i) ***Primary minerals:*** Primary minerals are the substances that are formed from primary igneous rocks via original crystallization of molten magma. Thus, the primary minerals form the original component of primary/igneous rocks. Examples: quartz, feldspars, hornblende, mica, etc. These minerals are originated from crystallization of the original molten mass of the earth and thus cannot be altered chemically. This is the mineral which is found in the soil but not formed in the soil. These are anhydrous in nature, and most prominent in sand and silt fractions of the soil.

Characteristics of Important primary minerals:

Quartz

Quartz is composed of Silicon and oxygen having continuous skeleton of Si-O tetrahedral layer with each oxygen being shared by two tetrahedral giving the formula of SiO_2. Quartz in its pure form is colourless and transparent or translucent.

- Quartz is very simple in its chemical composition i.e. an oxide of silicon (SiO_2).
- It has a specific gravity between 2.6 to 2.7 and can polarize any light passes through it.
- Most resistant to weathering as its structure is very closely packed that may prevents any sort of substitution.
- Quartz is colourless or transparent or translucent and has a vitreous luster i.e. it interacts with light in the same way that glass does.
- It occurs in mainly in granite (Igneous rock), Sandstone, Shale (Sedimentary rocks) and Schist, gneiss, quartzite (Metamorphic rocks)
- In most of the soils, its abundance is next to feldspars.

Feldspars (K and/or Na, Ca Feldspar)

- Earth crusts constitute 60 % of the feldspar mineral thus making it the most abundant mineral.
- These are the group of tectosilicate minerals (3-D framework of Si-tetrahedral)
- Feldspars are sub-divided into two groups- i.e. orthoclase and plagioclase.
- Orthoclase feldspar constitute potash feldspar, e.g. Microlite ($K_2O.Al_2O_3.6SiO_2$ or $KAlSi_3O_8$).
- Structure of majority of feldspars is more open than quartz which makes them less dense.
- Feldspars minerals are non-ferromagnesian minerals and act as source of Na, K, Ca and many trace elements in soils.
- Plagioclase feldspars are soda, calcium/lime or soda-lime feldspars, e.g. Albite- soda feldspar ($Na_2O.Al_2O_3.6SiO_2$); Anorthite- calcium feldspar ($CaO.Al_2O_3.6SiO_2$) and Oligoclase- soda-lime feldspar ($Na_2O.CaO.Al_2O_3.6SiO_2$).
- Feldspar is raw material generally used in glass making industry, ceramics, and in paint, plastics, and rubber as a filler and extender.

Pyroxenes & Amphiboles

- Pyroxenes are the group of inosilicates (single or double chain) minerals found in many igneous and metamorphic rocks.
- These are also termed as *ferro-magnesian* minerals as they contain mainly Fe, Mg and Ca in their structures, however, a small amount of Mn and Zn is also present.

- Pyroxenes have the general formula- $R_2(Si_2O_6)$ where R stands for mainly Ca, Mg and Fe. Examples: augite $CaO.2(Mg,Fe)O.(Al,Fe)_2O_3.3SiO_2$; diopside $Ca(Mg,Fe)(SiO_3)_2$; hypersthene $(Mg, Fe)SiO_3$ and enstatite $(MgSiO_3)$.
- Amphiboles are chemically more complicated; therefore, their chemical characterization is not simple as in case of pyroxenes.
- Amphiboles are easily weatherable and main source of micro-nutrients (Mn, B, Cu, Zn, Fe, Ni), Ca and Mg.

  Example: hornblende- $Ca_4(Na_2)(Mg,Fe)_{10}(Al,Fe)_2SiO_{14}O_{44}(OH),F)_4$.
- Pyroxenes are similar in colour, luster, and hardness to that of amphiboles but have slightly higher densities owing to the absence of hydroxyls.

Mica

- Mica is the groups of sheet silicates (phyllosilicates) minerals which are widely distributed in igneous, sedimentary and metamorphic rocks.
- Mica is highly reflective, and its peculiar cleavage property can split it into thin sheets.
- Mica is sub-divided into two groups: biotite and muscovite
- These minerals have plate like structure.
- Biotite, $K_2O.4\ (Mg, Fe)O.2(Al,Fe)_2O_3.6SiO_2$. is usually dark in colour.
- Biotite is richer in Fe and Mg than muscovite, $K_2O, 3(Al, Fe)_2O_3.6SiO_2$..

ii) ***Secondary minerals:*** The minerals formed due to weathering and recombination in primary rocks/minerals is called as secondary minerals. That means secondary minerals are formed when primary minerals undergo chemical or geological alterations such as weathering. Secondary minerals are formed and found in soil. Examples: clay minerals, gibbsite- $Al_2O_3.3H_2O$, limonite- FeO (OH). nH_2O, haematite-Fe_2O_3, etc. Generally, these minerals are hydrous unlike primary minerals. The secondary minerals dominate in the clay fraction and in some cases in the silt fraction of the soil.

Important secondary minerals:

i) ***Clay minerals***:

- Clay minerals are naturally occurring layered silicates e.g. illite, montmorillonite, chlorite, vermiculite, kaolinite, etc.
- These minerals contain Mg, Fe, and K in large amount along with O, OH, Al and Si.
- They are variable in colour depending upon their chemical composition.

- The structure of clay minerals is of sheet type like that of mica.
- In some clay minerals (montmorillonite, vermiculite- 2:1 type expanding type) structural sheets are loosely held together leading to swelling as the water molecules enter in their crystal lattice. Such clay minerals are termed as swelling-shrinkage type clay minerals. While in some clay minerals, structural sheets are firmly held together (1:1 type, 2:1 type) thus preventing the entry of water molecules in it and are referred to as non-expanding type minerals. The black cotton soils (*Vertisols*) of India have expanding type clay minerals; therefore, they are characterized by deep and wide cracks on the surface during dry conditions.
- Clay minerals are having net negative charge on their surface. Due to this negative charge, the cations are attracted which are not permanently bounded and may be exchanged with other cations in soil-water system. The amount of charge varies depending upon the type of clay mineral.
- Illite is the most common ly found clay minerals in almost all the soils (i.e. alluvial soils, forest soils, black soils, desert soils and red-lateritic soils)

Minerals are formed due to arrangement of silica tetrahedral and alumina octahedral positions inside it. A *tetrahedral* structure is formed by arranging three oxygen atoms closely packed in one plane and fourth atom on the interstice of three oxygen atom forming a void while *octahedral* structure is formed by arranging four oxygen atoms in the form of square and fifth and sixth oxygen atoms are placed in the interstice on the upper side and lower side of the plane.

Classification of silicate minerals

Major group	Structure	Chemical formula	Example	Structure
Nesosilicates	isolated silicon tetrahedra	[SiO4]4-	olivine	
Sorosilicates	double tetrahedra	[Si2O7]6-	epidote, melilite group	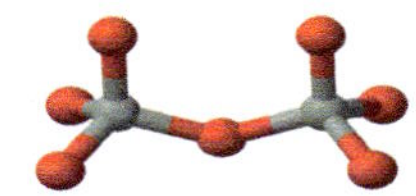
Cyclosilicates	rings	[SinO3n]2n-	tourmaline group	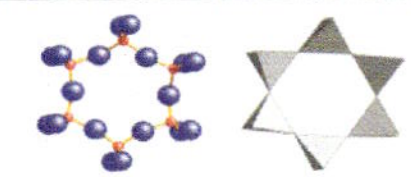
Inosilicates	single chain	[SinO3n]2n-	pyroxene group	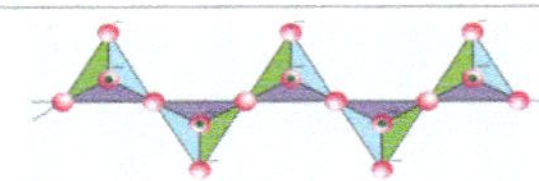
Inosilicates	double chain	[Si4nO11n]6n-	amphibole group	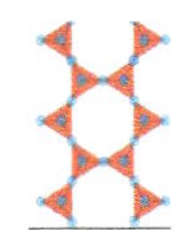
Phyllosilicates	Sheets/layers	[Si2nO5n]2n-	micas and clays	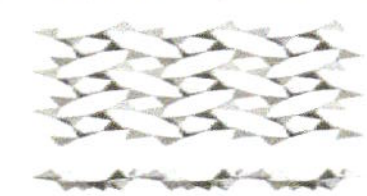
Tectosilicates	3D framework	[AlxSiyO(2x+2y)]x-	quartz, feldspars, zeolites	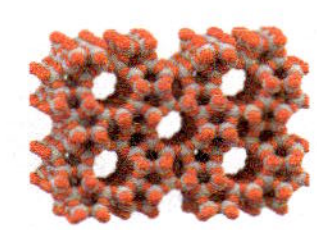

(Source: https://geologylearn.blogspot.com/2016/02/mineral-classication.html)

Haematite(Fe_2O_3)(Ferric oxide)

(Source: https://gotbooks.miracosta.edu/earth_science/chapter2.html)

- Haematite varies in colour from red to blackish red and has a reddish streak
- Metallic luster and hardness of about 5.
- Its presence in the rock is indicative of quick chemical change.
- Occurs as a coating on the sand particles and acts as a cementing agent.
- Haematite alters to limonite- FeO (OH).nH_2O, magnetite- FeO.Fe_2O_3,or
- Fe_2O_4 (ferrous-ferric oxide) pyrite-FeS_2, siderite- $FeCO_3$ and goethite, [FeO(OH).nH_2O]
- Paramagnetic nature and becomes strongly magnetic when heated.
- Generally haematite swells on absorbing water to form hydrated iron oxide.

Goethite **(FeO(OH)** ***i.e. Iron oxide-hydroxide*** **$Fe_2O_3.H_2O$**

(Source: https://gotbooks.miracosta.edu/earth_science/chapter2.html)

- Hydrated ferric oxide, yellow to brown in colour and present in abundance.
- Final product of most of the iron minerals, therefore, resistant to any further change except for absorption of water.
- An important colouring and cementing agent in soils.
- On dehydration, it forms haematite, and on reduction and carbonation, yields soluble iron.
- May be present in form of iron concretions.

Limonite **[$FeO(OH).nH_2O$]**

(Source: https://gotbooks.miracosta.edu/earth_science/chapter2.html)

- On absorption of more water goethite get changed tolimonite.
- Usually white but may also be pink or grey.
- Hardness is 5.3.

Gibbsite **($Al_2O_3.3H_2O$)**

(Source: https://gotbooks.miracosta.edu/earth_science/chapter2.html)

- Most common aluminium mineral in soils.
- White in colour hardness is 5.0-5.5; specific gravity 3.3-4.3.

- Abundantly found in highly weathered soils of the tropical zones, called *Laterites*(Oxisols).
- Its presence in soils shows the extreme degree of weathering and leaching under well drained conditions.

Calcite ($CaCO_3$)

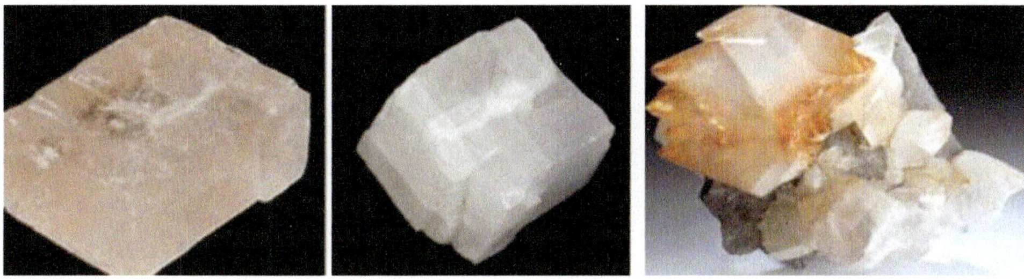

(Source: https://gotbooks.miracosta.edu/earth_science/chapter2.html)

- The calcite is formed when $Ca(OH)_2$combine with CO_2 or carbonic acid (H_2CO_3).

  $Ca\ (OH)_2 + CO_2 \rightarrow Ca\ CO_3$ (calcite) + H_2O

  $Ca\ (OH)_2 + H_2CO_3 \rightarrow Ca\ CO_3$ (calcite) + $2H_2O$

- White in ***colour*** with hardness 3.
- Widely distributed in sedimentary rocks which decomposes easily to soluble form as:

  $CaCO_3 + CO_2 + H_2O \rightarrow Ca\ (HCO_3)_2$(soluble form)

Dolomite [$Ca\ Mg\ (CO_3)_2$]

(Source: https://gotbooks.miracosta.edu/earth_science/chapter2.html)

- Dolomite is more resistant than calcite due tom its slow decomposition rate.
- It is less soluble in water than the calcite and thus is more resistant to erosion.
- Major source of Mg in soils.

Siderite ($FeCO_3$)

(Source: https://gotbooks.miracosta.edu/earth_science/chapter2.html)

- Siderite is a mineral of ferrous carbonate ($FeCO_3$) having a white streak and a vitreous luster or pearly luster.
- A weathered product of iron-bearing minerals of hardness 4 and alters to hematite or limonite.
- An important mineral in waterlogged soils.
- Siderite has been used as an iron ore and for steel production.

Gypsum ($CaSO_4.2H_2O$)

(Source: https://gotbooks.miracosta.edu/earth_science/chapter2.html)

- Gypsum is a mineral of sulphate group.
- Common mineral in desert soils and sedimentary rocks of hardness 2.
- Slightly soluble in water and is not easily leached.
- Precipitates as fine, powdery mycellium where the groundwater is rich in Ca^{2+} and SO_4^{2-}.
- Gypsum is used as an amendment to reclaim sodic soils and acts as a source of Ca and S for plants.

Apatite (Rock phosphate) $Ca_3(PO_4)_2$

(Source: https://gotbooks.miracosta.edu/earth_science/chapter2.html)

- ◈ Primary source of phosphorus in soils.
- ◈ Hardness is 5.
- ◈ Apatite decomposes readily under the influence of carbonic acid.
- ◈ Becomes immobile in soils as easily combines with clay, Fe-Al hydrous oxides, calcium carbonate, and become unavailable to plants.

2. ***Classification of minerals based on abundance:*** Minerals are classified into two groups according to their amount in the rocks.

   i) *Essential minerals:* The minerals which constitute major part of a rock and are regarded as the characteristic components of that rock. Examples: feldspars, amphiboles, pyroxenes, quartz and micas. These minerals are made up of silicon-oxygen atoms in various combinations, therefore, also known as silicate minerals.

   ii) *Accessory mineral:* The minerals which occur only in minor fraction and their presence do not affect the characteristics of the rock. They are not composed of silicates. They include phosphate, sulphate, oxides of iron and aluminium, e.g., apatite, tourmaline, magnetite, limonite, pyrite, calcite, gypsum, gibbsite, etc. These minerals are very important source of several plant nutrients.

3. ***Classification of minerals based on specific gravity:*** The minerals are classified into *light* and *heavy* minerals depending on their specific gravity.

   i) *Light minerals:* The minerals having specific gravity less than 2.85. Examples: quartz (2.60), feldspar (2.65) and muscovite (2.50-2.75).

   ii) *Heavy minerals:* The mineral having specific gravity more than 2.85. Examples: haematite (5.3), pyrite (5.0), limonite (3.8) and olivine (3.5).

4. ***Classification of minerals based on chemical composition:*** The minerals are grouped into several categories depending upon their chemical composition.

   i) *Oxides and hydroxides*: They are mainly anhydrous oxides of Si, Al, Fe, etc. but also include their hydrous oxides. The common examples are: quartz (SiO_2), haematite (Fe_2O_3), gibbsite (Al_2O_3), etc.

   ii) *Carbonates:* They are mainly carbonates of Ca, Mg, Fe, etc. Examples: calcite ($CaCO_3$), dolomite [$CaMg(CO_3)_2$].

   iii) *Sulphides:* These minerals are compounds of sulphide of heavy metals like Fe, Cu, etc., Example: Iron pyrite (FeS_2) - fool's gold due to its golden colour, Chalcocite Cu_2S, Chalcopyrite [$CuFe(S)_2$]. CuS_2copperdisulphide.

iv) *Sulphates:* These are sulphate bearing minerals, e.g., gypsum ($CaSO_4.2H_2O$).

v) *Phosphates:* These minerals are formed as tri-phosphates of calcium in combination with F, Cl, or OH, e.g., Fluorapatite [$Ca_5(PO_4)_3F$], Chlorapatite [$Ca_5(PO_4)_3Cl$], hydroxylapatite [$Ca_5(PO_4)_3OH,Cl$].

vi) *Silicates:* The minerals made up of silicon-oxygen atoms in various combinations are known as silicate minerals. Examples: feldspars, amphiboles, pyroxenes, quartz and mica.

vii) *Native elements*: The elements like sulphur, copper, etc. are native elements.

Chapter 3

Weathering of Rocks and Minerals

Rocks and minerals are weathered through series of stages to form soil. The processes involved in weathering leads to

- Formation of unconsolidated material from consolidated bed rocks through weathering or breakdown of the material
- Reorganization of these components though series of process to form soil

So, weathering refers to natural disintegration (reduction in size) and decomposition (chemical alteration) of the exposed rocks and minerals (bed rocks) due to actions of water, wind, air, temperature and biological agents. This process is a continuous process which over years leads to the formation of soil. First the rocks are broken down into smaller rocks and the mineral therein are exposed and changed into new minerals through either alteration or decomposition

Weathering of rocks is of two types: physical and chemical weathering. The permeability of rocks determines the rate of weathering of rocks.

For eg. sandstone a highly porous coarse textured rock weathers more fastly than a rocks with very fine texture for eg. Basalt. The minerals synthesized include 1) silicate clays 2) non silicate minerals like Iron and aluminum oxides.

1. **Physical:** The disintegration or break down of consolidated rocks into smaller pieces/fragments without changing its chemical composition or without producing any new substance is called physical weathering. It is caused by the mechanical force of water, air, temperature and biological agents.

   i) ***Physical weathering by water:*** Water causes physical weathering in many ways:

ii) Change in climatic condition leads to alternate freezing & thawing and wetting and drying. Thus when water present in cavities of the rock gets frozen, it expands (about 9% v/v) and creates a mechanical force within the rocks which breaks apart the rock. It is hydrothermal process which is more prominent in regions of cold climate. Some minerals expand or contract due to alternate wetting and drying condition may lead to swelling and shrinkage leading to disintegration of rocks.

iii) Falling rain drops and hailstorms beat the surface of rocks can cause abrasions on the surface of the rock thus leading to break down of rocks from the surface.

iv) Moving water has a tremendous carrying and cutting capacity. Water moving at a speed of 15, 30, 120 and 900 cm/sec can carry away fine sand, gravel, stones, and boulders, respectively. If the amount of suspended material is higher, the rate of disintegration of rocks will also be higher. Therefore, the weathering of rocks i.e. its disintegration is higher near the origin (start point) of river than the point where it finishes where it slows down steadily before meeting the sea.

v) Water due to its erosive action can remove weathered parts of rocks and exposes fresh surfaces to weathering.

vi) v) Glaciers during their movement cause cutting and crushing of the rocks.

*a) **Physical weathering by wind**:* Like water, wind also plays mechanical role in weathering:

i) Wind laden with fine materials (fine sand and silt) causes abrasive action on rocks and ultimately breaks down the rocks.

ii) Wind in combination with ocean waves cause weathering along the sea coasts.

*b) **Physical weathering by temperature:***

i) Temperature does have an influential effect on weathering of rocks and minerals. Sudden and wide changes in temperature greatly influence the disintegration/ breakdown of rocks. The rocks get heated due to the sun during day time, leads to expansion of rocks. Cooling and contraction of rocks occur due to fall of temperature at night. This alternate expansion and contraction due to diurnal (day-night) changes in temperature in hot and desert regions weakens the surface of rocks and disintegrates them.

ii) The minerals within a rock also vary in their rate of expansion and contraction. Example: The expansion of quartz is twice to that of feldspars.

iii) Colour of the rock also influences the rate of physical weathering of the rocks and minerals. The dark coloured rocks undergo quick changes in temperature due to more absorption of heat as compared to the lighter coloured rocks. The outer surface of the rock is comparatively darker in colour than the inner part of the rock. So gets heated faster than the inner portion, and thus gets expanded more than the inner surface. This differential expansion of minerals in a rock surface generates stress between the heated surfaces and the cooled subsurface, resulting in fractured/ break down of the rocks. The rock surface get heated and cooled quickly which may cause the surface layer to peel/strip off from the parent mass and disintegrate from the rock. This phenomenon of peeling off the surface heated rock from the parent rock due to variation in heat absorption is known as ***exfoliation***.

c) ***Physical weathering by biological agents:*** Biological agents such as plants (root penetration), animals (burrowing action and movement, particularly of larger animals) and cultivation by human also results in physical weathering.

- ◈ The pressure exerted by roots of higher plants on entering the cracks of the rocks causes disintegration or break down of bigger rocks into smaller ones i.e. physical weathering of the rocks.
- ◈ Plant roots also serve as food for microbes which act as biological agents for weathering.

a) ***Disruptive or abrasive action of plants***:

- ◈ Keep the rock moist when covered with vegetation may also lead to break down of rocks into smaller fragments
- ◈ Penetration of roots into cracks of rocks

b) ***Animals by burrowing action cause weathering:*** The animals like rabbits and rats burrow in the weak parts of rocks leading to physical weathering.

c) ***Activities of man:*** Activities like cultivation also cause physical weathering of rocks and minerals by breaking of rocks and minerals in to smaller pieces.

2. **Chemical:** The chemical weathering is a process of decomposition in which the original rocks and minerals are transformed into a new compound differing in chemical composition and physical properties.

- ◈ Surface area of the rocks exposed to environment may increase or decrease the chemical weathering effectively as most of the reactions take place on surface of the rocks and minerals. Thus, the smaller the particles of the rocks larger will be the specific area per unit volume of the rocks for chemical reaction to takes place.

- High chemical weathering resulted due to mineralogical composition of rocks. For example, quartz (SiO_2) responds very slowly to the chemical reaction due to its compact structure than olivine $(Fe,Mg)_2SiO_4$ because of their loose structure.
- The rate of chemical reaction increases with increase in CO_2 and other solvents in water, and temperature.
- The chemical weathering and its intensity is greater in tropical areas where climatic conditions are highly favourable due to availability of more water and warm temperature conditions while in deserts due to scarcity of water and cold temperate regions due to extremely low temperature, the rate of chemical weathering is low.
- Plants, animals and other organisms also contribute to chemical weathering. Bio-organism influences the weathering by producing carbon dioxide and certain organic acids on decomposition of organic matter that react with rock constituents. Bacteria, fungi and actinomycetes in rocks and minerals produce appreciable amounts of organic acids which activate the solution and cause the chemical weathering of the rocks and minerals. Production of HNO_3 by nitrifying bacteria can cause chemical weathering.

Processes of chemical weathering: The chemical weathering is brought about by solution, hydration, hydrolysis, oxidation, reduction, carbonation, etc.

1. ***Solution***: Water is a universal solvent which dissolves the constituents present in the rocks. These dissolved constituents (NaCl) are removed continuously from the rocks due to flowing/percolating water due to which rocks no longer becomes solid and developed holes in it which can be disintegrated and decomposed easily. The action of solution is increased considerably due to acidification of water by the dissolution of organic and inorganic acids.

$$NaCl + H_2O \longrightarrow Na^+, Cl^-, H_2O$$

(soluble salt) *(dissolved ions surrounded by H_2O molecules)*

2. ***Hydration***: Hydration is the chemical attachment of water molecule with a rock or mineral. Primary minerals are unhydrous in nature which may get hydrated when comes in contact to moist humid conditions. The hydration is always accompanied with increase in the volume of rock/mineral. Hydrated minerals become soft and more readily to weathering. Minerals like feldspars, amphiboles, micas become hydrated and form hydrous compounds. Example:

$$Fe_2O_3 + 3H_2O \longrightarrow Fe_2O_3.3H_2O$$

(Haematite, red-black) ***(Limonite, yellow)***

$$Al_2O_3 + 3H_2O \longrightarrow Al_2O_3.3H_2O$$

(Bauxite) *(Gibbsite)*

$$CaSO_4 + 2H_2O \longrightarrow CaSO_4.2H_2O$$

(Calcium sulphate) *(Gypsum)*

3. ***Hydrolysis:*** Hydrolysis is a double decomposition process. It involves the splitting of water into H and OH ions, and reaction of substances with water to form hydroxidesand other new substances that are usually more soluble than the original material. Water, thus, acts like a weak acid. Example:

$$KAl\,Si_3O_8 + HOH \longrightarrow HAl\,Si_3O_8 + KOH$$

(Orthoclase) ***(Acid silicate)***

The hydroxides in the presence of CO_2 change into CO_3 and HCO_3.

4. ***Oxidation:*** Oxidation is the process of absorption of oxygen to the minerals with the loss of electrons. It is an important chemical reaction occurring in well aerated environment where oxygen supply is high and biological demand is low. Oxidation process is particularly important in iron containing rocks and minerals as iron is easily oxidizeable. In most of the primary minerals, iron is present in ferrous (Fe^{2+}) form. On oxidation, it changes to haematite. Moisture present in air and water helps in oxidation of rocks and minerals.

$$4FeO + O_2 \longrightarrow 2\,Fe_2O_3 \text{ (Haematite)}$$

(Ferrous oxide) ***(Ferric oxide)***

$$4Fe_3O_4 + O_2 \longrightarrow 6Fe_2O_3$$

(Magnetite) **(Haematite)**

Rocks containing ferromagnesian silicates such as pyroxenes, hornblende, biotite and chlorite are susceptible to oxidation. Similarly, minerals containing manganese and sulphide group are also susceptible to oxidation.

5. ***Reduction:*** Reduction is the reciprocal of oxidation i.e., removal of oxygen from the mineral under water saturated conditions (such as below water table) where oxygen supply is low with the gain of electrons. The net effect of these conditions is to reduce a metal, e.g. ferric oxide to ferrous oxide which is highly soluble and mobile. If ferrous iron persists in the system, it tends to form sulphides (ferrous sulphide, FeS_2) and other ferrous compounds such as cheleopyrite, $CuFe(S_2)$.

6. ***Carbonation:*** It is the combination of CO_2 with bases. Carbonic acid, although a weak acid of CO_2 but an important agent in chemical weathering of rocks and minerals as it increases the solubility of minerals. The decomposition of organic matter liberates CO_2. Carbonation tends to produce carbonates and bicarbonates:

$$H_2O + CO_2 \longrightarrow H_2CO_3$$

(Carbonic acid)

$$CaCO_3 + H_2CO_3 \longrightarrow Ca\ (HCO_3)_2$$

(Calcite) (Calcium bicarbonate)

The solubility of Ca $(HCO_3)_2$ is considerably higher than calcite. In soils, H^+ ions present on clays are also available for such reactions. Example:

$$Ca\ Al\ Si_2\ O_8 + H\ \boxed{\text{Micelle}} \longrightarrow H_2AlSi_2O_6 - Ca\ \boxed{\text{Micelle}}$$

(Anorthite) (Acid silicate)

The H^+ of acid clay replace the bases of the minerals.

Chapter 4

Soil Formation

Formation of soil is the process of transforming rocks and minerals through weathering into soil. Soil formation starts with the weathering of rocks and minerals to form regolith which is finally converted to soil due to different factors and processes acting on to it. The processes of weathering of rocks and minerals are primarily destructive process resulting in formation of unconsolidated mass from the consolidated material (Bed rock) called *parent material*. However, the processes of formation of soil are constructive process resulting in formation of soil profile.

The soil formation can be shown as:

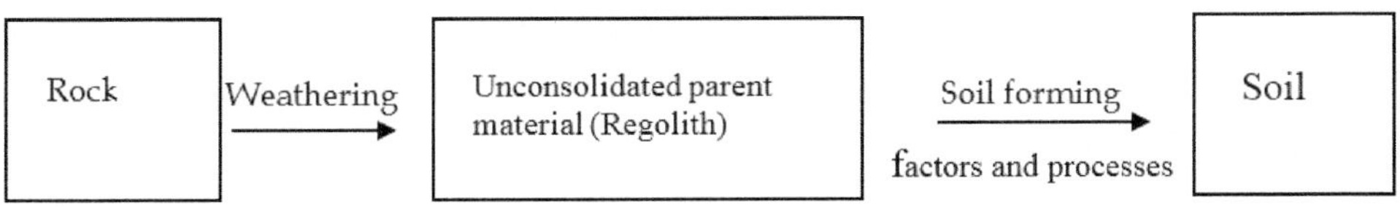

SOIL FORMING FACTORS

Docuchaiev (1900), Father of Soil Science, develop the relationship of different factors on the formation of soil and expressed in the form of equation as:

$$S = f(p, cl, o)$$

Where S = soil, p = parent material, o = organism

Later, *Hans Jenny* (1941) formulated the combined influence of all the soil forming factor in developing the profile and expressed it in the form of equation as:

$$S = f(c, b, r, p, t, \ldots)$$

Where S = any soil property (texture, colour, permeability, pH, etc.), c- climate (temperature and rainfall), r – relief of the land/ topography, p – parent material, t – time of formation of soil i.e. age of the soil.

All of these factors depend on each other in the formation of soil. Each factor has its own significance in development of soil but not in equal magnitude. Some factors are more effective in determining the nature of soils under specific conditions like climate and biosphere.

Thus, Jenny stated that any soil property developed due to weathering of rock and minerals is a function of all the five soil forming factors. **Joffe (1949)** divided these factors into two groups, viz. ***Active factors and Passive factors.***

Active: ***Active soil forming factors are those that supply energy which acts on the regolith or the parent material for soil formation.*** They act as a catalyst for its formation. These factors include climate and biosphere.

Active factors represent the driving forces acting upon the parent material for soil formation and promote the processes causing changes in soil properties.

1. ***Climate:*** Climate is the most relevant and significant factor that affects formation of soil directly through precipitation and temperature, and indirectly through activities of flora and fauna. Depending upon the amount of precipitation, temperature and evapo-transpiration the climatic regions are classified as:

   i) *Arid:* The precipitation is less than the evapo-transpiration and soils remain dry almost throughout the year.

   ii) *Humid:* The precipitation exceeds than the evapo-transpiration. The excess water leads to leaching of basics salts (mainly) which is followed by the translocation of colloidal clays in lower depths in soil profile.

   iii) *Temperate:* The region with cold and humid environment in winters with warm and humid summers.

   iv) Sub-Tropical and tropical: It includes warm to hot temperature with humid moisture regimes.

a) ***Effects of rainfall/precipitation on soil formation:*** Rainfall influence the process of soil formation in several ways.

   i) Higher level of rainfall/moisture favours *hydration and hydrolysis* of rocks and minerals causing rapid weathering. It also provides conditions for better growth of vegetation and microorganism.

   ii) High precipitation favours the *leaching and percolation* processes. Thus, soils of humid regions carry away the soluble constituents due to leaching of basic cations (Ca, Na, Mg, K), and therefore, becomes acidic in nature (low pH). Percolating water will favours more profile development with distinct horizonation.

iii) Well drained and well developed soils of humid regions are characterized by iron hydroxides in various stages of hydration, therefore, the soils of such region vary in their colour from grey to brown to dark brown depending on the organic matter.

iv) High rainfall will favour more vegetative growth due to increase of H^+ and nitrogen concentration in the soil.

v) In high rainfall regions, the clay content of the soil increases due to weathering and mobilization, thus characterized by high cation exchange capacity.

vi) In low rainfall areas, soils of arid and semi-arid regions have high amount of soluble salts because of less leaching and high temperature. Therefore, soils are generally alkaline in nature (high pH) and have accumulation of carbonates of Ca and/or Mg at shallow depths.

vii) Hot and arid region soils possess low cation and anion exchange capacity because of low organic matter and clay content, and thus have less profile development. Low rainfall and high temperature build up the concentration of basic cations mainly Ca, Mg, Na and K.

b) ***Effect of temperature on soil formation:*** Mean annual soil temperature has significant influence on its formation being highest in the regions of equator and declines gradually as we moves towards the poles.

- The radiations from sun striking the surface of the earth are being controlled by certain factors of which vegetation is the most important. Vegetation acts the buffer on soil temperature as it reflects 90% of the incoming solar radiations.

- The amount of radiations reaching the earth surface is largely determined by daily (diurnal) or seasonal fluctuations. The heat moves inwards in the soil during the day due to warming and start returning back to the surface as soil surface cools down during night. These diurnal changes are more in dry arid regions but are observed up to a 30 cm depth of the soil from surface. Below 30 cm depth, temperature of the soil may become uniform and constant. Therefore, soil temperature is considered to be true at 50 cm depth. That is why well water is cooler in summer and warmer in winter.

- The sun rays falling straight on to the earth's surface heat up the soil more (as at equator) than that falling at an inclined angle (away from the equator). That is why when sun is at right angled (mid of the days) to the earth surface, more radiations will reach the surface as compared to when it is in morning or in evening.

- Temperature can affect weathering up to a considerable depth. In warm humid climate, rocks and minerals can weather to a much greater depth, forming a deeper solum than in cold temperate region where depth of soil is shallow due to less intense weathering.

The effect of temperature in its rate of soil formation can be related as:

i) **Vant Hoff's Law** (1884) which states that "for every 10^{0}C rise in temperature, the rate of chemical reaction increases by 2-3 times". This applies for the weathering of rocks and minerals. Many chemical reactions are more active in relatively higher temperature areas than in lower temperature regions.

ii) **Ramann (Ramann's weathering factor)** expresses chemical weathering as a hydrolytic decomposition of the silicates. As temperature increases, the relative degree of dissociation of water increases which enhances the chemical weathering of rocks and minerals leading to faster formation of soil.

iii) Organic matter content of soil decreases as the annual temperature rises. Humus cannot build up in soils of regions having temperature above 25^{0}C due to fast decomposition of organic matter. In soils with temperature less than 10^{0}C, organic matter can be greater than 20% due to poor microbial activities/decomposition.

It may be concluded that rainfall and temperature together will influence the soil formation by increasing the rate of reactions viz. chemical and biological leading to the formation of soil. Thus, climate has a direct influence on soil formation through providing water and heat that may interact with rock material and indirectly through activities of flora and fauna by releasing energy in the form of organic acids that reacts with rocks and minerals for soil formation. In tropical regions, high rainfall and temperature will leach the soluble solutes to lower soil depth, thus the soils become acidic. The soils formed in humid temperate regions are greyish in colour which often modifies to black or brown depending upon the amount and nature of organic matter and iron oxides. In tropical regions, soil derived specially from igneous and metamorphic rocks is yellow and dark red coloured.

2. ***Biosphere:*** Different microorganisms, plants, animals and human beings all influence the formation of soil and in modifying its properties.

   i) *Vegetation* influence soil formation by providing the quantity and nature of organic matter. Soil developed under forest areas has more horizons with non-humified organic material than those of the soils developed under grassland regions. Grassland has fibrous root system which decomposes readily in situ thus favours the accumulation of organic matter in the rooting zone.

ii) *Vegetation* also influences the translocation and accumulation of mineral substances. Translocation is more in forest soils than in grasslands since because coniferous trees favours the acidic conditions due to leaching of basic cations thus decreasing the activity of microorganism and in-turn organic matter decomposition. Grassland regions experienced low rainfall and neutral to alkaline condition, thus favours the retention of mineral substances in the soil. Thus, organic matter of the grassland on decomposition produces organic acids which may speed up the rate of weathering of rocks and minerals. Decomposition of organic matter may result in release of essential minerals in soil and it also helps in improving the nutrient and water holding capacity of the soil.

iii) *Microorganisms* including micro-flora and micro-fauna play an important role in biochemical transformation. They decompose soil organic matter which is dominant in forest and grass lands as compared to deserts. Thus, microorganisms lead to increase the organic matter decomposition, humification, nitrogen fixation, ammonification, nitrification, denitrification, oxidation, reduction etc.

iv) *Macroorganism* such as burrowing animals, rodents, earthworms, ants and termites are important in soil formation as they mix soil within soil profile. They may also help in decomposing the minerals and rocks by producing the digestive juices.

v) *Human-beings* through their land use activities cause both deteriorating and beneficial effects on soil formation.

Passive: Passive soil forming factors are those that represent the source of soil forming material and the conditions acting upon it. These factors provide the sources on which active soil forming factors work for development of soil. These factors are parent material, relief (topography) and time (age).

I) *Parent material:* Parent materials are those substances from which the soils are formed due to climate and biosphere. It determines the development of soil profile and properties of soils. *The parent material is the unconsolidated chemically weathered mineral matter and organic materials from which the soil is developed by pedogenic processes.*

The rate of soil formation will depend on the mineralogical nature, composition and size of particles of parent material.

1. ***Nature and composition of parent material:*** Some minerals undergo weathering easily while others are resistant to the weathering. The composition of minerals present in the parent material largely in-fluenced the formation of soil. If the parent material is composed of

quartz, feldspars, etc (resistant to weathering), the time taken for development of soil will be long and the soil formed will be coarse in texture. If the parent material is composed of amphiboles, pyroxenes, olivine, calcite, etc., i.e. less resistant to weathering, the time taken will be short and the soil formed will be fine in texture. The order of resistance of minerals to weathering is: **Quartz > K-Feldspar > Na Ca-Feldspar >Biotite> Hornblende > Amphiboles > Pyroxenes > Olivine (ortho-silicate) > Calcite > Gypsum.**

2. *Size of particles of the parent material:* The size of particles influences the soil formation by controlling the rate of percolating water through the parent material.
   i) If the parent material is composed of coarse particles (large sized), water will pass quickly through the parent material causing less chemical changes.
   ii) If the parent material is composed of fine particles (small sized), less water will enter and pass through the parent material resulting again in less chemical changes.
   iii) However, if the parent material is composed of medium sized particles, optimum amount of water enters and available within the parent material which results in high chemical changes in parent material leading to fast development of soil.

Parent materials influencing the formation of soils: Different parent materials have different effect on profile development which may yield different soils.

i) Soil texture is being influence by parent material. Highly siliceous parent material such as sandstone may give sandy soils while basic rock and fine-grained sediments tends to give fine textured soils.
ii) Soils formed under sandstone parent material although have deeper soil with very low fertility while those develop under limestones, and alluvium parent materials forms fertile soils.
iii) Acidic igneous rocks (granite) usually weather slowly and results in light textured (coarse) sandy soils (podzolic soils) with low base saturation and fertility.
iv) Basic igneous rocks (basalt, rich in Ca, Mg, etc.) weather rapidly giving rise to fertile and fine textured soils, e.g., black cotton soil (Vertisols) in central India.
v) Sedimentary rocks formed from basic igneous rocks on weathering forms fine textured soils with high levels of bases. Soils with montmorillonite clay minerals with high fertility are formed from these parent materials.

vi) Hard pure limestone yields sandy shallow soils whereas impure soft limestone yields to deeper and finer textured soils.

Parent material is not only the factor that affects the properties of the soil. Different kind of parent material may give rise to comparable soils if climate and vegetation of an area are identical or the same kind of rocks / parent material can result in different soils, if climate and vegetation varies. Eg: Basalt may give rise to highly weathered red soils in humid tropics and black cotton soils (vertisols) in semi-arid and sub-humid/sub-tropics. Initially soil properties are mainly determined by parent material but at later stages of development, influence of parent material diminishes.

Geological classification of soil parent material

Two groups of inorganic parent material have been recognized

1. Sedentary (formed in place i.e. in situ)
2. The soils formed from the parent material through transported to other places from the place of their origin and re-deposited some-where else.

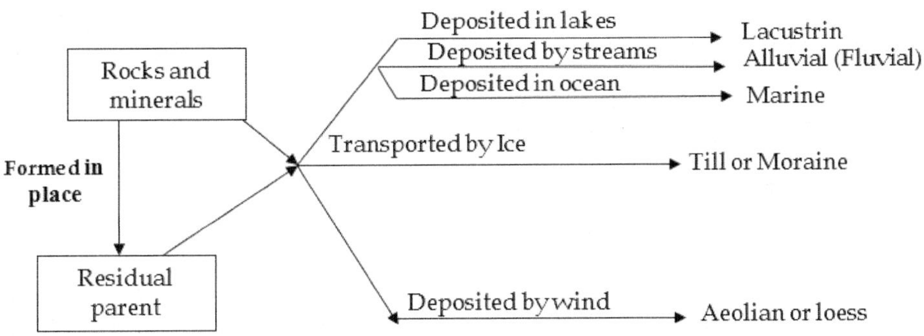

Different parent material (transported and deposited)

Residual parent material: It is develop in place from underlying rocks (Bed rocks). Typically, it is more intensely weathered. These soils are comparatively low in Ca and Mg concentration due to leaching out of bases from the solum. Red and yellow colors develops due to formation of oxides of iron, are the characteristics features of the soils formed due to residual parent material in hot-humid areas whereas, oxidation and hydration of iron may be hardly noticeable in cooler and drier climate. In warm humid climate, it is more oxidized and well leached.

***Alluvium or alluvial soils*:** The soil material transported and subsequently deposited by the action of water is known as alluvium and soils formed are called as *alluvial soils*. Alluvial soils are considered to be fertile soils since they carry the sediments from the surface of the soils to other places.

Colluvium or colluvial soils: The material transported and deposited from the top of slopes and carried downwards by *gravity* is known as colluvium and soils are called as *colluvial soils.* The colluvium materials are poorly sorted formed near the base of steep slopes. A soil developed from the colluvium material is dependent on the sources of the material and are generally not preferred for agriculture purposes because of their unfavorable physical and chemical characteristics.

Lacustrine deposits consist of materials that have settled in the stagnant water of lakes. The sediments carried by streams are deposited in lakes, oceans, seas etc. The coarser fragments settles near the bank of river while the finer particles to a greater distance. Over the period of time these deposits becomes quite deep due to deposition of sediments in layers which are subjected to weathering giving rise to some productive agricultural soils. They may vary in texture from sandy to clayey.

Aeolian or Loess: The materials blown by the wind are known as ***loess*** if texture is silty and ***aeolian*** when deposits are primarily sand. *Aeolian* deposits are light and porous and are generally less productive than the soils formed by *loess.*

Glacial drift consists of the materials which is picked up, mixed, disintegrated, transported and accumulated by the action of glacial ice and/or water resulting from melting of glaciers. The materials deposited by the melting glaciers vary widely in particle size and are called ***till*** or ***moraine.***

II) ***Relief or topography (Synonyms terms):*** Relief refers to the difference in elevation between the highest and lowest part of an area measured in terms of slope or slope angle. Topography is the shape of the land. An area's topography may be flat or slopy, hilly or mountainous. Topography of an area includes the area's elevation, relief and landforms. The relief or topography refers to the configuration of the land surface.

The topographic designations as per FAO guidelines (1990) are:

Land surface	Slope
Flat to almost flat	0-2%
Gently undulated	2-5%
Undulating	5-10%
Rolling	10-15%
Hilly	15-30%
Steeply dissected	>30% with moderate range of elevation (<300m)
Mountainous	>30 % with great range of elevation (>300m)

Soil formation on flat to almost flat topography: On level topography, water received through rain percolates through the soil. Under such conditions, the soils formed may have soil profiles of distinct horizons. In general, level lands with small gradient (difference in elevation between the two points) often have impaired drainage conditions that retard the development of soil profile.

Soil formation on steep hills/undulating topography: With the identical rainfall and comparable parent material, the shallow soils formed on top of hills, and with indistinct horizons and low profile development. This may be due to

i) Surface soil material carried down to the slope as soon as it is formed.

ii) Reduce percolation of water due to loss of water in the form of surface runoff due to slope

iii) Limited water available for the plant growth

The steep slopes and accelerated erosion often promotes downward movement of soil under the influence of gravity. This process is called as *solufuxion* or *soil creep*.

Soil formation in low lying areas (depression): The low lying areas have high moist conditions than the soils on level topography due to the water falling as runoff from nearby areas. High moist condition due to runoff water leads to high vegetative growth and slow down the decay of organic matter resulting in the formation of soils rich in organic matter (*Mollisols*).

The depression areas in *humid regions* are relatively wet and remain wet for longer time which favours the development of organic (peat or muck) soils. Such soils are finer in texture and have poor drainage conditions. If those areas carry the salts from nearby areas, the soils in the low lying area may become salt marshy with cultivation of salt tolerant crops or salts may affect the plants up to the toxic level too. When soils in the low lying soils are highly acidic, iron get deposited in low lying areas to form the limonite (bog iron).

Red soils are formed on granite rocks at higher topographic positions; the black cotton soils are developed on alluvium derived largely from basalt rocks at lower topographic positions.

The soils that are formed from the same parent material but differ in their topographic positions within a given landscape shows differential profile development, the sequence of such soils is termed as ***toposequence***.

Soil formation and exposure: Topography affects soil formation by affecting temperature and vegetation through slope exposures (aspect). In the northern hemisphere, soils on south and west facing slopes receive more direct sun-rays and are comparatively warm and dry than north and east facing slopes. In arid climates, these drier south and west facing slopes are often less productive than the soils on north and east facing slopes. The opposite slopes are affected in the southern hemisphere. In cool and moist regions, these warmer sites may be highly productive. Higher temperature on south and west slopes results in greater loss of water by evaporation. The soils on southern exposures (facing the sun) are warmer with high fluctuations in temperature and moisture. The northern exposures, on the other hand, are cooler and more humid. The eastern and western exposures occupy intermediate positions in this respect.

III) *Time:* Time for the formation of soils is very slow process which takes hundreds to thousands years to develop a completely mature soil having distinct horizonation. "The period taken by a given soil from the stage of weathering to the stage of maturity is considered as time". With the term mature we means, the soil having fully developed A, B & C horizon. The time period required for the formation of soil or horizon depends upon climatic condition of the area, type of the parent material and relief/topography of the region etc. When the soil conditions are favourable, a soil profile with distinct horizons may develop within a period of around 200 years whereas in less favorable circumstances, soil development may take thousands of years. The age of the soil can be estimated by *radio carbon dating, pollen analysis or carbonate carbon*, etc.

All soils do not develop at the same time or age. The development will be faster in warm and humid climatic conditions and forested conditions in case of permeable and loose parent material. Extreme climatic conditions (hot, dry and cold), impermeable and consolidated parent material, and very steep slopes delay in formation of soils.

There are five stages of soil development (Mohr and van Baren):

1. *Initial stage:* Initial stage of soil formation indicates that there is an un-weathered parent material.
2. *Juvenile stage:* Initial stage of the weathering which is just initiated but still many of material is unweathered
3. *Virile:* Majority of the readily weatherable minerals has decomposed, the percentage of clay has considerably increased and some development is noticeable. The soil material less susceptible to weathering is still found in considerable quantities.
4. *Senile:* At this stage, decomposition is almost completed, and only the highly resistant minerals to weathering are still unweathered.
5. *Final:* Soil development has been completed under prevailing environments.

The term *young soil* means the soil where the processes of formation of the soil are still in operation and all properties of soil are still changing in the profile and soil forming factors and processes do not have a great impact on the soil profile. A *mature soil* indicates that a parent material is fully weathered and there is no further change observed in properties of the soils (no change of parent material) with environment. In completely mature soil, profile with horizonation is distinct.

In young soil, clay from primary materials is still being formed whereas in mature or old soil, the clay is more or less in equilibrium with the primary minerals. In young soil, more amount of clay is observed at surface as compared to lower layers whereas in case of mature soil, percentage of clay was found higher at subsurface soils as compared to surface.

In India, soils in the alluvial regions are classified on the basis of age of alluvium as ***Khadar*** (young) and ***Bhangar*** (old).

Interaction of Soil Forming Factors: The formation of soil is a diverse and complex process which is brought about by the five major factors working in combination. For instances, soil with good drainage, mild temperature and high rainfall will probably support plenty of plant life. Plants, in turn, decompose, producing carbon dioxide which combines with water from rainfall to form carbonic acid. The resulting acidity increases the solubility of parent material causes leaching of Na, K, Ca, and Mg, and the soil becomes more acidic. High rainfall also translocates some clays and organic colloids deeper into the B-horizons.

The same climate and parent material might produce waterlogged soil in a low lying valley or depression instead of a gentle slope (different topography). The poor drainage results in stagnant water and non-aerated soil, often with poor plant growth and decreased rates of organic matter decomposition. The accumulating drainage water also lead to the accumulation of dissolved salts. The horizons of these soils, developed from common factors but differing topography, will become very different.

The regions of high rainfall and good drainage develop acid soils having surface horizon (A) with high organic matter, and removal of soil colloids from the surface and their accumulation in the deeper horizon (B). If the soils have been developing for a long time, they tend to have high clay content, well differentiated horizons, acidity in wet climates, and salt accumulation in some soils of arid regions.

Chapter 5

Processes of Soil Formation

The pedogenic soil forming processes are highly complex and dynamic processes involving various biological and physico- chemical reactions which may takes place simultaneously with each other. The pedogenic processes change the unconsolidated mass into soil with distinct horizonation. There are two main kinds of horizons

i) Organic (O)

ii) Mineral (A, B and C)

The O-horizon is rarely observed in arid and semi-arid soils due to low moisture conditions. Wetness generally favours the increasing thickness of O-horizon. Mineral horizon has low content of organic matter and overlying the bed rock.

There are generally two categories of pedogenic processes involved in the formation of well-developed soil i.e., *fundamental/basic soil forming processes and specific soil forming processes.*

I) Fundamental processes: The basic/fundamental processes involved in soil formation are that include transformation and translocation are:

1. ***Humification:*** Humification is the process of decomposition of organic matter and synthesis of new organic substances i.e. the process of formation of humus from decomposition of raw organic material. This is a highly complex process involving decomposition through various micro and macro fauna activities. The decomposition of organic material takes place in two phases:

   - ***Mineralization:*** "It is a biochemical breakdown of dead plant tissues by soil microorganisms to produce a simple structured soluble organic substances, mineral compound, metal cation and gases".

- ***Humification:*** "Soluble organic substances rearrange themselves into large molecules by polymerization and become insoluble". Humification leads to the formation of humus which provides the charged sites for the retention of cations on the humus. Other major product of humification is the production of polysaccharides which help in binding the nutrient ions and the soil particles.

Mechanism of decomposition of organic matter

The dead plant leaves and stem after falling on the soil are attacked by microorganisms which cause the breakdown of different plant compounds. In this process of decomposition the simpler compounds are decomposed at first, such as starches & sugars. After that the more resistant material like proteins, cellulose, are decomposed and then most resistant material are decomposed such as tannins, and finally dark coloured amorphous substance known as *humus* is formed. In contrast, the decomposition of the leaves of deciduous trees by earth worms and bacteria is much faster and may be completed within a year, leaving little organic matter at the surface.

Microorganisms control oxidation–reduction potential in soils. Under aerobic conditions, the liberation of CO_2, H_2O and energy results in high oxidation potential of soils (biological oxidation) which results in loss of electron and decrease in the pH of the soil whereas under anaerobic soils, there is a gain of electron and increase in pH of the soils resulting in reduction of Ferric (Fe^{3+}) and Manganic (Mn^{4+}) to Ferrous (Fe^{2+}) and Manganous (Mn^{2+}) which are toxic for plant growth and also promotes denitrification.

2. ***Eluviation:*** "Eluviation (washing out) is a process of removal of soil constituents (clay, Fe_2O_3, Al_2O_3, SiO_2, humus, $CaCO_3$, other salts in suspension or solution) by the percolating water from the upper to lower layers in the soil". The upper layer where loss of soil constituents occurred is called as *eluvial* layer or A-horizon or E-horizon. The lower layer, where soil constituents accumulate is termed as illuvial layer or B-horizon. Such a horizon is more pronounced in podzols or spodosols. This mobilization and translocation of soil constituents results in the difference in texture between the two layers.

Leaching: "Leaching is defined as the translocation and removal of soil constituents or ions from the entire solum in solution form".

Eluviation and leaching of an element depends on its relatively mobility and rate of water percolating through the soil. In arid regions, the highly mobile elements (SO_4^{2-}, Cl^-) may retained in soil solum or move to the surface mainly due to high evapo-transpiration rate especially under shallow groundwater conditions. In semi-arid regions, the highly mobile elements leached to lower layers whereas less mobile element may concentrate to surface soil upto some depth in the form of carbonates of Ca and Mg. In humid regions, leaching of the cations mainly basic occurs due to high rainfall.

"The mechanical movement of clay and iron oxides from A horizon without undergoing chemical alteration is called Lessivage". Under grassland vegetation, there is a rapid recycling of minerals due to dense root system and preferential liking of grass roots for cations whereas under deciduous forest, acids condition will develop which will reduce the activity of the microorganisms, due to which decomposition of organic matter and production of polysaccharides that binds the particles slows down. Thus the water stability of peds is reduced. As a result dispersion of clays begins and translocation in suspension begins and consequently a clay depleted alluvial horizon develops.

3. ***Illuviation:*** "The process of deposition of soil materials (removed from the eluvial horizon) in the lower layer is called as *illuviation*". The horizons where the accumulation takes place by this process are termed as *illuvial* horizons (B-horizon). The process leads to difference of texture between E and B_t horizons, and higher fine clay in the B_t horizon.

II) Specific processes: The specific soil forming processes are:

1. ***Podzolization:***It is the process of soil formation resulting in the formation of podzols (spodosols) and podzolic soils.

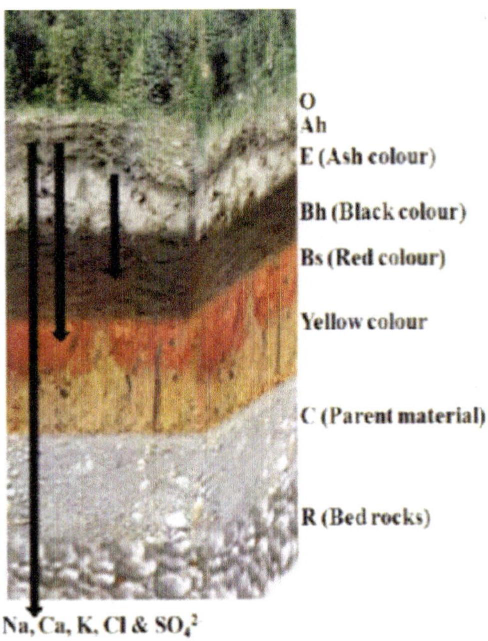

Source: https://www.researchgate. net publication/314453325_ SOIL _GENESIS_ AND_CLASSIFICATION

It is a process of translocation of iron and aluminium oxides, organic material and clay from the A-horizon to the B-horizon and also the removal of bases from the entire solum, producing an acid soil with whitish grey *silicious* A-horizon and yellow to brown compact and impervious B horizon.

In many respect, podzolization is the negative of calcification. The process of calcification tends to concentrate $CaCO_3$ and $MgCO_3$ in lower B-horizon whereas podzolization leaches the entire solum of Ca and Mg carbonates. Apart from Ca & Mg, other bases are also removed.

The favourable conditions for podzolization are:

i) **Climate:** A cool with humid climatic condition is highly favourable for formation of podzols. Under this condition, leaching removes the cations as soon as they are released by weathering. Podzolization can also occur in tropics at high altitude but because of the dominance of other pedogenic process, it prevents the completion of podzolization process.

ii) **Parent material:** Parent material with silica as the chief constituent element with poor stocks of weatherable minerals favours the process of podzolization as it facilitates translocation of water through the soils.

iii) **Vegetation:** Coniferous pines and hemlock vegetation on decomposition produces acid such type of vegetation favours podzolization.

iv) **Leaching and translocation of sesquioxides:** Under acidic condition, organic matter decomposition produces various organic acids thus reducing the microbial activity and results in declining polysaccharides production. The organic material combine with sesquioxides and the clay minerals, resulting in formation of org-sesquioxide and org-clay complexes, which gets solubilized and thus percolate with water to the lower B-horizon. Further, hydrolysis of aluminium ions makes the soil more acidic due to release of H^+.

$$2Al + 6(H\text{-}OH) \longrightarrow 2Al(OH)_3 + 6H^+$$

As the Fe and Al move out, the upper A-horizon showed a greyish bleached coloured horizon. Acids produced also dissolve and remove the carbonates from the mineral layer resulting in lowering of soil pH to 4.0 to 4.5. Under such pH, the solubility of iron and aluminium is optimum and are removed from the upper mineral layer/horizon.

Land Use: Low in fertility and are mainly used for forestry and grazing. It may also be used for growing oats, potato and clover.

2. ***Laterization (Gr. Later means brick):*** The laterites refers specially to those cemented soils which when dried becomes hard and used as bricks for construction of buildings.

Laterization is the soil formation process in which silica is translocated out of the solum, leaving behind sesquioxides that remained in the upper layer. The alumina and ferric oxides become hydrated and accumulate (to extent of 70-80 % of total mass) in the soil together with quartz. Such soils become hard like a brick when dried and are called the *laterites* or *latosol.*

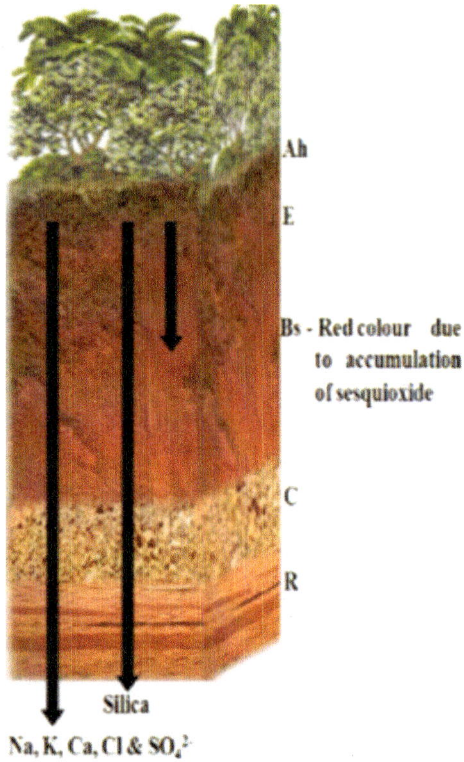

(Source: publication/314453325_ SOIL _GENESIS_AND_CLASSIFICATION)

The conditions leading to laterization are:

i) **Climate:** This process is most favouable in warm and humid (tropical) climate with rainfall ranging from 2000-2500 mm and continuous high temperature (>25°C) throughout the year. In tropical regions, warm and humid climate causes intense weathering of rocks and minerals producing silicic acid, aluminium and iron hydroxide or hydrated oxides of iron and aluminum by hydrolysis and oxidation.

ii) **Natural vegetation:** The tropical rain forests are most favouable for the process. The organic remains of such forest are incorporated in the soil very slowly because of high biological oxidation owing to high temperature. Because of pronounced leaching, organic matter and soluble minerals are removed rapidly.

iii) **Parent material:** Basic parent materials, with considerable amount of ferromagnesian minerals (pyroxenes, amphiboles, biotite and chlorite) which on weathering releases Fe, are favourable for the formation of lateritic soils. The iron released during weathering is oxidized to its oxide components which coats the clay, silt ans sand particles imparting grayish colour to the soil.

iv) **Soluble bases:** Calcium, magnesium, sodium and potassium are quickly released in initial stages upon the hydrolysis of minerals. Neutral and even alkaline conditions resulted in the upper layer due to mobility of these cations and pH becomes near to 7. In neutral and alkaline solution, solubility of silica is encouraged and is removed from silicate minerals. Iron, aluminium and manganese oxides are left behind together with quartz.

Laterization in conjunction with podzolization has resulted in the development of six zonal soils in the forested warm temperate and tropical regions. These are: Laterite, Redish brown laterite, Yellowish brown laterite, Red podzol, Yellow podzolic soils and Terra rossa

Characteristics:

i) Deeply weathered soils (Deep B horizon)

ii) Non-plastic and non-cohesive and have granular structure

iii) Low CEC and very low in fertility due to lack of organic matter and hydrous clay

iv) High fixation of P by Fe and Al oxides

v) Soil crusting is the another problem of this soil

Land use:

i) Shifting agriculture

ii) Low intensity grazing

iii) Plantation of cashewnuts, coconuts, coffee, banana, pineapples with adequate nutrients

3. ***Gleization:*** Glei (*Russian word- mean – blue-grey or green clay*). This is the soil formation process which led to the development of gleiz i.e. gley horizon in the subsoil of the profile due to poor drainage condition (waterlogging, lack of O_2). Such soils are called *Hydromorphic* soils.

The poor drainage may be due to

i) Depression area (low land)

ii) Imperviable soil material

iii) Deprived of aeration

Under this process, ferric iron reduces to ferrous form and their hydroxides dominate which is responsible for blue to greyish colours of the horizon. Where the groundwater fluctuates, the grey horizon shows distinct mottling of yellow or rusty brown colours due to oxidation and reduction phenomenon.

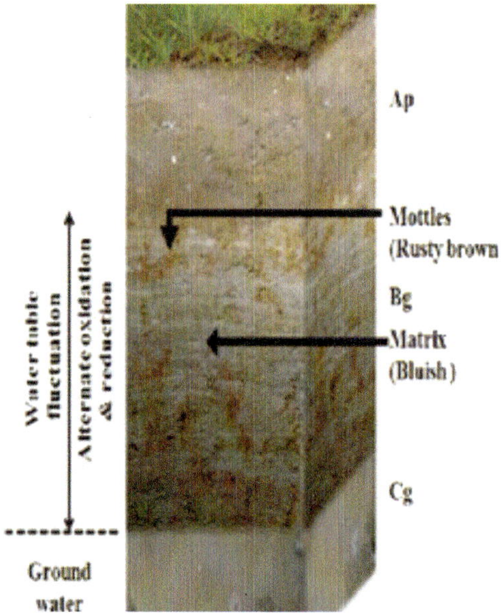

Source: publication/314453325_ SOIL _GENESIS_AND_CLASSIFICATION

Land use: It is mainly used for growing rice however, if drainage condition are improved then theses soils can be used for growing wheat, sorghum, millets etc.

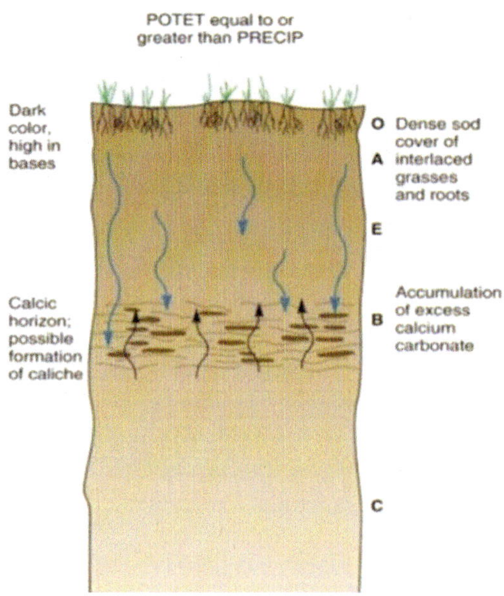

(source:https://www.studyprobe.in/ 2023 /06/soil-formation-calcification.html)

4. ***Calcification:*** Calcification involves deposition and precipitation of calcium carbonate ($CaCO_3$) or magnesium carbonate ($MgCO_3$) in

some part of the soil profile approximately to the depth to which water percolates. Calcification process results in arid and semi-arid regions due to restricted rainfall. Rainfall is insufficient to leach much of the calcium and divalent cations from the soil profile which results in the redistribution of Ca and Mg carbonates. Surface soil has sufficient calcium to maintain the soil colloids in a high degree of base saturation.

The accumulation of $CaCO_3$ may result in the development of *calcic* horizon. Calcium is readily soluble in acid soil and /or when CO_2 concentration (CO_2 liberated from plant root or microbial activity) is high in root zone as:

$$CO_2 + H_2O \longrightarrow H_2CO_3$$

$$H_2CO_3 + Ca \longrightarrow \underset{\text{(soluble)}}{Ca(HCO_3)_2}$$

But again due to lack of microbial activity and moisture due to prolonged dry weather, the process is reversed and results in the precipitation of $CaCO_3$.

$$Ca(HCO_3)_2 \xrightarrow[\text{Dec. in } CO_2]{\text{Inc. in Temperature}} \underbrace{CaCO_3}_{(precipitates)} + H_2O + CO_2$$

The process of precipitation of $CaCO_3$ after reaction is called *calcification* and resulting illuviated horizon of carbonates of calcium which is known as *B_k horizon*.

5. ***Decalcification:*** It is the removal of $CaCO_3$ or calcium ions from the soil by leaching.

$$\underset{(insoluble)}{CaCO_3 + CO_2 + H_2O} \xrightarrow[\text{Inc. in } CO_2]{\text{Dec. in Temperature}} \underset{(soluble)}{Ca(HCO_3)_2}$$

6. ***Salinization*** **(Solonchaks):** Salinization is the process of soil formation in which sulphates and chlorides of calcium, magnesium, sodium and potassium in soils accumulates in the form of salty (*salic*) horizon. The electrical conductivity of the saturation extract of the soil is >4 dsm^{-1}. Saline soils are commonly observed in arid and semi-arid regions. Due to high evaporation rate, upward movement (through capillary rise) of these salts takes place which results in accumulation of the salts in surface layers. Salt accumulation may also result from irrigation or seepage in areas of impeded drainage. The following conditions may accumulate salts:

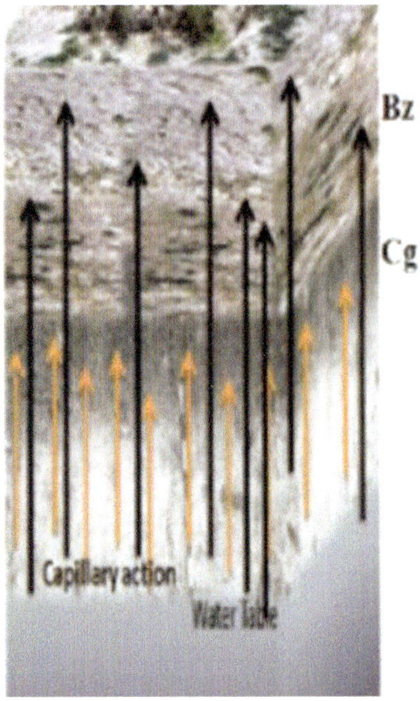

Source: publication/314453325_ SOIL _GENESIS_AND_CLASSIFICATION

i) Arid and semi-arid regions

ii) Lower topographic position iii) Imperfect drainage condition

iii) Deposits of salts along the sea coast

iv) Irrigation water with high salinity

Soil Characteristics: Soils have more water soluble Ca and Mg salts due to which pH seldom rises 8.5. Soils have granular structure due to flocculation caused by soluble salts. However, high salt concentration inhibits plant growth. Such soils can be reclaimed by introduction of surface and subsurface drainage systems so as to lower down the EC < 4 dS/m. The soils with high pH and ESP on exchangeable complex are *sodic* or *alkali* soils.

7. ***Desalinization:*** "The removal of excess soluble salts from horizons or soil profiles by flooding the soils with water of low salinity and also improving the drainage conditions by installing artificial surface or sub-surface drainage systems".

8. ***Solonization or alkalization:*** "The process of accumulation of Na^+ ions on the exchange sites of the soil resulting in the formation of sodic/alkali soils (*Solonetz*), is called as solonization". In sodic soils, the exchangeable sodium percentage (ESP) of soil increases to a value > 15. The soil colloids both organic and mineral become dispersed due to presence of Na^+ and tend to move downwards. The dispersion results in poor physical conditions and this forms black organic-clay coatings on the ped surface. Thus, these soils are also called as black alkali soils.

The sodic soils are dark brown to black colour with very low permeability to air and water, friable surface horizon, a hard and impermeable subsurface due to illuviated clays and have typical columnar structure. Such soils are commonly observed in Indo-gangetic plains. These soils are reclaimed by removal of Na from the exchange complex of the soil and saturating it with Ca.

9. ***Solodization or dealkalization:*** "The process refers to the removal of Na^+ from the exchangeable sites with Ca^{2+} ions which cause flocculation of soil is called as solodization" . The Ca^{2+} and /or Mg^{2+} ions are concentrated in the water which leads to the flocculation of the soil mass and also used to leach the solonetz. These ions result in flocculation of clay colloids. The dealkalization process is shown as:

$$2NaX + CaSO_4 \longrightarrow Na_2SO_4 + CaX$$

(leachable)

10. **Pedoturbation**: It is the process of mixing or turning of soil. It may be-

   i) *Faunal Pedoturbation*: It is the turning of soil by the soil fauna such as ants, earthworms, rodents and man

   ii) *Floral pedoturbation*: It is the turning of the soil by soil flora and its roots as in tree tipping that forms pits and mounds

   iii) *Argillipedoturbation*: It is the mixing of material in the soil by churning process caused due to swelling and shrinkage of clays Eg. Deep black cotton soils of central India

   iv) *Cryopedoturbation*: It is the mixing of the soil material due to churning caused by frost like that occurs in Gelisol soil order.

Chapter 6

Soil Morphology

Soil morphology is the describing the soil entity as a whole, its appearance, characteristic features and general properties as expressed in the profile of the soil. The morphology is expressed by number, thickness and arrangement of different horizons. "A soil horizon is a layer of soil approximately parallel to surface with characteristics produced by soil forming processes".

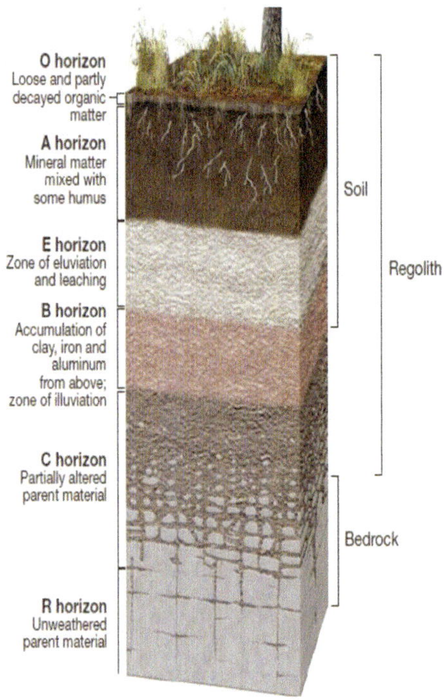

Schematic representation of a soil profile

Source: www.digitallylearn.com

Pedon: (Introduced by USDA Soil Survey staff) "It is the smallest volume of the soil that can be recognized as soil individual and has the smallest area for which we should describe and sample the soil to represents the nature and arrangements of its horizon and variability in other properties".

Polypedon: A soil body consisting of more than one pedon is termed as polypedon.

The most conspicuous part of any soil is its surface. There is a continuous transfer of energy and matter from soil surface to atmosphere and vice versa. The surface of the soil may be leveled or uneven, it may consists of granular or crusted soil and cracked, friable or hard, vegetated or fallow, mulched or exposed. Such surface conditions affect the processes of exchange and interaction of radiations and heat, water and solute movement, and diffusion of gases. Thus, it is important to study the surface of the soil but simultaneously surface soil characteristics cannot described the characteristics of the soil profile as a whole. For characterizing the soil throughout from depth, it becomes necessary to explore the soil from surface up to a certain depth which can be done by digging a trench.

"A vertical section of the soil from the soil surface to the parent material is called a *soil profile*". In deep soils, it may be up to a depth of 1.5 -2.0 m while in shallow soils, it is up to the bed rock/parent material or up to water table in the case of waterlogged soils.

The soil profile consists of a sequence of distinct layers, in terms of properties viz. physical, chemical and biological, called soil *horizons*. Maximum five master soil horizons may be observed in a soil profile. These are represented by capital letters- O, A, E, B and C. The characteristics of the soil horizons are:

O-horizon: It is upper most organic horizon above the mineral soil which is composed of organic material derived from plant residues and dead animals. This horizon is usually found in soils developed under forests and generally absent in cropland and grassland.

A-horizon: A-horizon is the topmost mineral horizon mixed with humus which lies at or near the soil surface constituting around 1% organic carbon under agricultural soils. Due to considerable organic matter, A-horizon is generally darker than the horizons underlying it. This horizon consists of major biological activities and the very fertile zone/layer of soil but also easily eroded with water and wind.

E horizon: The E horizon lies below the A-horizon. It is the horizon from which few soil materials such as silicate clays, carbonates and bicarbonates, oxides and hydroxides are continuously removed by percolating water. It is also called as a zone of *eluviation* and usually found in forest areas. In soils developed under cropland and grassland where horizon E is absent, the colloidal soil materials viz. silicate clays, oxides of Fe and Al and carbonates of Ca and Mg are translocated continuously from the A horizon through percolating water to B horizon.

B horizon: The B horizon lies under the surface A-horizon/E- horizon which constitutes less organic material than overlying A-horizon. This horizon is also known as the zone of *illuviation* where the materials viz. colloidal clays, organic matter etc. leached from the above lying A-horizon/E-horizon accumulates. The thickness of mineral B-horizon is more than the mineral A-horizon. The A-horizon is more porous than the underlying B-horizon as A horizon is having more organic matter content than B horizon that may leads to more aggregation and thus increase the porosity. Also, the pressure exerted by A horizon on to the B horizon compacts the lower horizon due to which porosity decreases. In certain soils, B horizon becomes so dense that may inhibit gaseous exchange, water drainage and restrict the penetration of roots.

C horizon: The C-horizon lies below the B-horizon which consists of weathered rock also called parent rocks. If the soil is formed from the bedrock *in situ*, the C horizon consists of a weathered and fragmented rock material. Otherwise, the C-horizon may also contains the materials carried due to water (alluvial), wind (Aeolian) or Ice (glacial deposits).

In young soils where the soils have been formed in the recent time, the B-horizon is generally not formed or it may be absent or indistinct but in a matured soil i.e. fully matured soils, A- and B-horizons are more visible/distinct and sometimes due to completion of weathering processes, the C- horizon may disappear. In recent alluvial soils the profile differentiation is hardly noticeable. The horizons O, A, E, and B are collectively termed as solum (true soil) whereas horizons O, A, E, B and C are collectively termed as regolith.

The soil acts as a favourable medium for the growth of the plants and to do so it must have the capability to retain and move water and nutrients from soil to root and from roots towards shoots and also has the capability to degrade the toxic chemical substances to make them in toxic. Thus, soil fertility may be defined as "the inherent capacity of a soil to supply nutrients to plants in adequate amounts and in suitable proportional" where as "the ability of a soil to produce a crop is known as *soil productivity*". Productive soils are always fertile but chemically fertile soils may or may not be productive. A soil which contains good amount of nutrients elements required by plants in an available form preferred by plants, but does not possess high acidity or alkalinity and is free from toxic agents, are considered to be fertile soils but it may not ensure the growth of the plant. It rather depends more on the physical structure of the soils i.e. physical fertility which includes movements of water and nutrients translocation, air transport and heat conduction. Therefore, physical as well as chemical fertility are equally important for higher productivity of soil. Examples of soils which are chemically but not physically fertile are:

i) Soil are considered deserts if water is absent inspite of being rich in all other essential plant nutrients,

ii) Lack or absence of oxygen in the depression low lands where soils are enriched with nutrients carried by moving water from nearby areas and deposited in the low lands but the land becomes infertile if flooded for a longer period of time, and

iii) Nutrients and water of the sub-soil remain unapproachable to the crops if a dense plough layer restricts their root penetration. Also, the soil fertility depends on its bulb density. Compacted soil easily obstructs root penetration. Thus, roots under this condition can be able to reach nutrients and becomes unproductive.

iv) Soils enriched with nutrients may not be productive until that nutrients is not converted into its available form required by plants. This may be due to low microbial activity and less nutrient recycling. Soil flora and fauna play a vital role in creating good soil structures by producing channel and pores, concentrating tiny soil particles into stable aggregates and fragmenting and mixing organic substances throughout the soil.

v) Low productivity of the soil might also be clue to high soluble salt concentrations inspite of its hight fertility. Salts may lead to poor passage of water & thus nutrients.

vi) Soil productivity may also be related to pH of the soil. Very high and very low pH may restrict the availability of nutrients or may lead to nutrient toxicity.

Chapter 7

Soil as a Disperse three Phase System

The ultimate disintegrated product of rocks is called *soil mechanical separates* or *primary particles*. Plant and animal remains and residues, left on the soil as organic matter, are decomposed by micro-organisms to form *humus*. Humus and other mineral-binding agents, such as calcium carbonate, oxides of iron and aluminum, and clays, etc., bind soil separates to form *secondary soil particles* called *aggregates*. Both, primary and secondary particles form a soil matrix, in the pores of which both air and water move and are retained.

The soil, thus, formed acts as the basic substrate of life on the earth. It thus acts as a source for biological activities beside being as a medium for plant growth. Soil also helps in recycling the waste products otherwise it might poison the soil system. Soil is also used for construction purpose i.e. in building earthen dams, building roads etc.

SOIL COMPONENTS AND PHASES

The soil is divided into four major components and it constitutes three phases. The four components include

i) Inorganic solids,

ii) Organic solids,

iii) Water, and

iv) Air.

An inorganic component includes solid portion of the soils which is primarily made up of minerals matter. Organic components are thus which constitutes organic matter of the soil i.e. made up of plants residues and dead remains of animals. The water component of the soils consists of an aqueous soil solution existing within soil compounds. The gaseous component is the soil air which constitutes gases like nitrogen, oxygen, carbon dioxide, methane, nitrous oxide, and other inert gases. When the conditions are optimum for plant growth and in a medium textured surface soils, the solid soil components comprises of inorganic mineral matter as well as organic matter constituting 50% of the total volume of the soil of which 45% constitutes soil solids (mineral fraction of the soil) and 5% organic matter, while liquid and gases comprise 25% each. The organic fraction of most mineral soils constitutes 5% or less. The air and water component of the soil varies with the environmental conditions. The water component may occupy the entire pore space of the soil when it rain heavily or during flood irrigation. In that situation, the air component vanishes to almost zero. However, after the period of few days of rainfall or irrigation, half of the pores spaces are occupied with air and half still occupied by water. Under completely dry condition, water in the pores can be occupied by air.

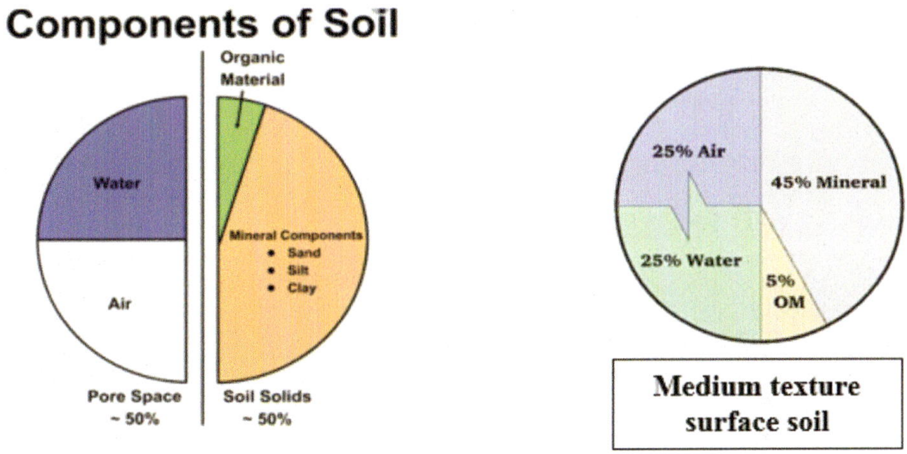

(Source: *https://www.ctahr.hawaii.edu/mauisoil/a_comp.aspx*)

For engineering purpose, the solid components should be of 80–90% of the total volume of the soil. These components of the soil varied in their mass, volume, spatial and energy status due to both natural and management factors.

Table: General properties of components and phases of soil

Phases	Components	**Composition**	**Properties**
Solid	Inorganic	Products of weathering-quartz, feldspars, hornblende, pyrite, calcite, gypsum, apatite, alumino-silicates, etc.	Skeleton, matrix ρ_s=2.6–2.7 Mg m^{-3}
	Organic	Remains of plants and animals, Living organisms, usually <5% in mineral soils	Large surface area, very active, p_s=1.1 to 1.3 Mg m^{-3}

Phases	Components	Composition	Properties
Liquid	Soil solution	Liquid with ionic compounds (Na, K, Ca, Mg, Cl, NO_3, PO_4, SO_4)	Heterogeneous, dynamic, discontinuous ρ_w=1.0 Mg m^{-3}
Gas	Soil air	N_2, O_2, CO_2, CH_4, C_2H_6, H_2S, N_2O, NO	ρ_a=1–1.5 kg m^{-3}, variable, dynamic
ρ_s - practice density,		ρ_w - density of Water,	ρ_a density of air

THREE PHASE DISPERSE SYSTEM OF THE SOIL

Natural system can consist of one or more substances and one or more phase.

- The system which comprises of one substance only i.e. a homogenous system having constant and steady state physical properties. Such system is called as monophasic system. Eg. Block of ice
- The heterogeneous system may consist of single substance if that system exhibits differently in different phase of the system. Eg. Mixture of ice and water, which although constitutes uniform chemical properties but possess different physical properties because it exists as two different phases.
- A heterogeneous system consisting of different substances may also be monophasic Eg. a system of salt and water mixed together constitutes single phase.
- A system of many substances can be heterogeneous.

In a heterogeneous system, most of the physical properties such as phenomenon of adsorption of H_2O and nutrients ions, capillarity, ion-exchange, dispersion, flocculation of soil particles, aggregation, swelling & shrinkage, heat of wetting, etc., occur at the interface of two phases.

The three phases existing in the nature are solid, liquid and gas. Systems in which at least one of the phases is composed of numerous small sized particles are called *disperse systems*. Soil is a dispersed three-phase system as it is made up of solids, liquids and gases. The soil is a heterogeneous, polyphasic, particulate, disperse and porous system of which interfacial area per unit volume is enormously large. Other disperse systems are colloidal solutions, gels, emulsion.

Composition and importance of the phases in agriculture:

Solid phase: The matrix of the soil solid phase consists of inorganic (mineral) and organic particles. The inorganic fraction is composed of discrete rock and mineral particles varying in chemical and mineralogical composition, shape, size and orientation (crystalline structure). This solid phase may also contains amorphous (non-crystalline) materials such as that of minerals like imogolite and allophane (short ranged), hydrated oxides of iron and aluminum etc which are found less in quantity in general soils. Solid phase may also be composed of organic fraction consisting of the dead remains of flora and fauna in different stages of their decomposition Un-decomposed materials are roots, stems, leaves, etc., and well decomposed products are polysaccharides, humus, lipids, gums, etc.

This solid phase is important in following respects:

1. The physico-chemical reactions like ion exchange, adsorption, desorption, capillarity etc. takes places on solid phase of the soil i.e. solid phase provides the seat for almost all reactions.
2. The mineral particles of the solid phase joined together to form the pores spaces that facilitates the movement of air, water and nutrients and also helps in penetrating the roots and their proper growth in the soil.
3. The solid phase acts as a source of nutrients and provides mechanical support to the plants.
4. The organic matter of the solid phase helps in aggregating the soil particles held responsible for the loose and friable conditions of the soils.
5. Solid phase also helps in holding the water and making use of that water at the time of need by the plants. It provides energy for the activity of soil microorganisms.

Liquid phase: The liquid phase is the soil water which always contains dissolved substances combinedly called as soil solution.

1. Water held within the soil pores increases the availability of nutrients by making it in a soluble form.
2. The movement of nutrient ions from soil to root and from roots to leaves takes place through water.
3. Water meets the hydrogen requirement of the plants.
4. It is the essential for taking out all the biochemical processes occurring in the plants.
5. Water is required for managing the aeration of the soil, maintaining optimum temperature in soil and plant, and also the activities of microbes in the soil.

Gaseous phase: Gaseous phase constitutes soil air. Air in the soil is not distributed as a continuous phase but located in the some pockets between the pores spaces and its composition varies from one location in the soil to another. Different soil reactions and activities of fauna and flora greatly influence the content of soil air. Moisture content of soil air varies and can reach 100 % under water saturated soils. The CO_2 content of the soil is around 10 times higher than that of atmosphere air. In general, the soil CO_2 is several times more concentrated than the 0.03%, commonly found in the atmosphere air. The O_2 decreases accordingly and in extreme cases, it may not be more than 10-12% as compared to 20.6% of the atmosphere.

Soil air is important in the following respect:

1. Soil air is necessary as it facilitates oxygen which is required for the respiration of roots and soil micro-organisms.
2. The oxygen of the soil air transforms the mineral elements into the available forms, for example, nitrogen to nitrate ($N-NO_3$); sulphur to sulphate ($S-SO_4$), etc.
3. Nitrogen of the soil air is fixed in the soil by microbes and becomes available to plants.

VOLUME-MASS RELATIONSHIPS OF SOIL CONSTITUENTS

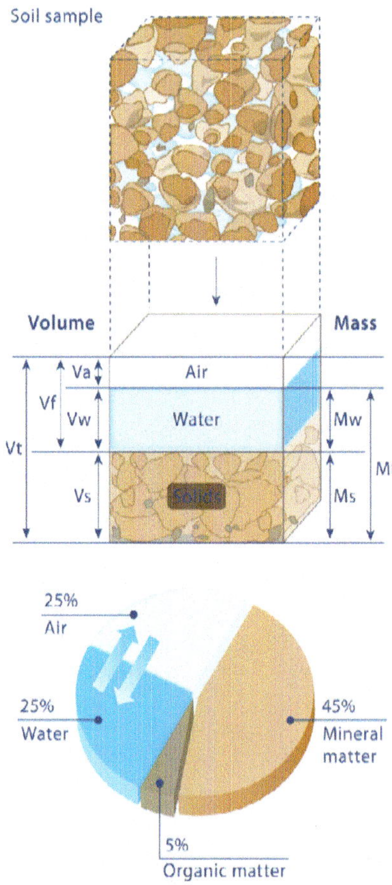

Schematic representation of soil as a three phase system

Source: https://gpambala.ac.in/wp-content/uploads/2018/10/Soil-Mechanics.pdf

In the schematic representation of three phase system of the soil, consider the volume and masses of the components of the soil system be represented as:

Volume of air = V_a,

Volume of water = V_w

Volume of solid = V_s,

Volume of pores ($V_a + V_w$) = V_f

Total volume of the soil ($V_a+V_w+V_s$) = V_t

Mass of air ≈ Zero

Mass of water = M_w

Mass of soil solid = M_s

Total mass of soil = M_t

The masses can be represented in term of their weight (w) as:

Weight (w) = mass (m) x acceleration gravity (g)

On the basis of the three phase of the soil system, we can defined and express the few quantitative relationships

1. Particle density (ρ_s)

Particle density may be defined as the mass of soil solid per unit volume of soil solid. It is generally expressed in terms of g cm^{-3} (CGS) or Mg m^{-3} (SI). In almost all the mineral soils the particle density may range from 2.6 to 2.7 Mg m^{-3} due to dominance of quartz mineral which has a density in this range in the coarser fraction of the soil. In finer fraction of the soil, even the soil possesses same density but the presence of Iron oxides and other heavy minerals may increase the density of the soil to 2.9 Mg m^{-3}.

Specific gravity (σ_g) can be expressed as the ratio of density of any material to the density of pure water at 4°C and atmospheric pressure or can be expressed with reference to some other material.

If the organic matter of the soil is < 1.0 %, then the particle density of the mineral soil is 2.65 Mg m^{-3}. But if the organic matter of the soil is more than 1.0 %, then the particle density of the soil can be calculated by the sum of the density of minerals multiplied by the fraction of minerals and density of organic matter multiplied by fraction of the organic matter.

$$\rho_s = \rho_m . f_m + \rho_o . f_o$$

where ρ_m & ρ_o= particle density of the mineral and organic matter, respectively

f_m & f_o = fraction of mineral and organic matter, respectively

FACTORS AFFECTING PARTICLE DENSITY OF THE SOIL

a) Chemical composition of mineral particles

The average ρ_s of the soil is 2.65 Mg m^{-3} due to dominance of quartz, feldspar, mica and aluminosilicate clay minerals but because of the presence of minerals like pyrite, hematite, limonite, magnetite etc. ρ_s of the soil may increase beyond 2.65 Mg m^{-3}.

b) Organic matter

Mass of the organic matter is much lower than the mass of equal volume of mineral particles. The ρ_o lies between 1.1 to 1.5 Mg m^{-3}.Thus, the organic matter of the soil decreases the particle density of the soil.

c) Depth of the soil

The particle density of the surface soil is less than that of subsurface soil since because surface soils have high organic matter content than subsurface.

2. Bulk density (ρ_b)

Bulk density is the ratio of mass of dry soil solid to the bulk volume of the soil. This bulk volume is the sum of volume occupied by soil solid and pore space.

Bulk density may be expressed as dry bulk density and wet bulk density. Dry bulk density is the ratio of dried particles to the total volume of the soil whereas wet bulk density is the mass of field soil sample per unit bulk volume of the soil sample. The mass of field soil sample is the mass of water plus the mass of soil solid.

$$\rho_b\,(\text{dry}) = \frac{M_s}{V_t}$$

$$\rho_b\,(\text{wet}) = \frac{M_t}{V_t} = \frac{M_s + M_w}{V_t}$$

The average bulk density of an ideal soil is 1.33 Mg m^{-3}.

$$\rho_b\,(\text{sandy soil}) = 1.2 \text{ to } 1.8 \text{ Mg m}^{-3}$$

$$\rho_b\,(\text{clayey soil}) = 1.0 \text{ to } 1.6 \text{ Mg m}^{-3}$$

Factors affecting bulk density of the soil

a) Texture

Fine textured soils have more total pore space than coarse textured soils. Thus, the mass per unit volume of fine textured soil is less than those of coarse textured soils. So, fine textured soils such as clay, clay loam, silt loam have lower bulk density than the coarse textured soils.

b) Structure

Well aggregated soils have lower bulk density than poorly aggregated soils. Thus, spheroidal type structured soils which possess more aggregation have lower bulk density than platy like structure. Bulk density is also affected by degree of compaction and swelling and shrinkage characteristics of the soil. Soils having high swelling and shrinkage clays have low bulk density than the compacted soils.

c) Organic matter

Organic matter and its decomposed products binds the primary soil particles to form soil aggregates resulting in an increase in the total pore spaces and decrease the mass per unit volume of the soil. Organic matter on decomposition produces polysaccharides that help in binding the particles to each other which in turn will promote more aggregation. Thus increasing the pore spaces and decreasing the bulk density of the soil.

d) Soil depth

Bulk density of surface soils is lower than subsurface soil. The higher bulk density of subsurface soils is due to lower content of organic matter, less aggregation, less root penetration and more compaction due to weight of overlying soil layers.

e) Nature of crop and soil management

Bulk density for the soil decreases due to grasses grown in place of cultivated crops as because the grasses have extensive root system which forms rounded crumbs or granules due to pressure exerted by growing roots.

Addition of crop residues and farm yard manures or compost may further promote aggregation leading to decrease the bulk density of the soil.

f) Tillage

Tillage may temporarily decrease the bulk density of the soil as it loosen the soil surface but intense tillage increase the bulk density of the soils due to reduction in organic matter and weakening of soil structure. In mechanized agriculture, the wheel of heavy machines use to pull the implements, apply amendments and harvest crop can yield limiting soil compaction (plow pans).

3. Porosity (f)

The porosity of the soil is that portion of the total soil volume not occupied by soil solid but occupied by air and/or water. It is expressed in percentage or fraction. It represents the relative pore space in the soil.

$$f = \frac{V_f}{V_t} \times 100 = \frac{V_a + V_w}{V_t} x\ 100$$

Porosity of the coarse textured (sandy) soils is less than in fine textured (Clayey) soils inspite of being larger individual pore size of the former. In fine textured clayey soils, the variable soil porosity is due to the nature of the clays i.e. the swelling and shrinking of the clays, formation of aggregates, disperse, capacity of forming cracks on drying etc.

Its value generally lies in the range of 0.3 – 0.6 (30-60 %)

Air filled porosity (f_a) = $\frac{V_a}{V_t}$

Air filled porosity indicates the aeration status of the soil. Critical limits of f_a = 0.1 (10%) below which the lack of air will negatively affect the growth of the plant.

Relationship of porosity (f) with bulk density (ρ_b) and particle density (ρ_s)

We know

$$V_t = V_s + V_f$$

$$V_f = V_t - V_s$$

The total volume of pore spaces per unit volume of soil

$$\text{i.e.} \frac{V_f}{V_t} = \frac{V_t - V_s}{V_t} = 1 - \frac{V_s}{V_t}$$

$\frac{V_s}{V_t}$ can be expressed by chain rule as $\frac{M_s}{V_t} x \frac{V_s}{M_s}$

So, $\frac{V_f}{V_t} = 1 - \frac{M_s}{V_t}\frac{V_s}{M_s}$

$$\frac{M_s}{V_t} = \rho_b; \qquad \frac{V_s}{M_s} = \frac{1}{\rho_s}$$

Therefore, $\frac{V_f}{V_t} = 1 - \frac{\rho_b}{\rho_s}$

Or f = $(1 - \frac{\rho_b}{\rho_s})$

Size of the pores

Pore spaces in the soil are divided into macropores and micropores. Macropores are the pores which have diameter greater than 0.06 mm. After saturation, water drains out from these pores by gravitational forces, as a result of which pores are unable to retain water and mostly filled with air. Macropores are also known as *aeration porosity or non-capillary porosity* which may be defined as the per cent of pores filled with air after the soil has drained to field capacity. Aeration porosity influences plant growth by influencing permeability of water and air. If aeration porosity is high, the soil has adequate aeration, high infiltration and permeability.

Micropores are also called as *capillary porosity* which is defined as the percentage of pore spaces occupied by capillary water after the soil has drained to field capacity. If the capillary porosity is high as in clayey soils, there may be inadequate aeration, low infiltration and permeability, high water retention at various suction values from field capacity to permanent wilting point.

4. Void ratio (e)

The void ratio represents the fractional pore space of the soil. It relates the pore space to the volume of soil solid rather than total soil.

$$e = \frac{V_f}{V_s} = \frac{V_a + V_w}{V_t - V_s}$$

It ranges between 0.3 and 2.0. The advantage of void ratio over porosity is that any change on pore volume affects only the numerator of the equation whereas in case of porosity such a change affects the numerator and denominator both. Void ratio is used by engineers whereas porosity is being preferred by agricultural scientist and engineer too..

Relationship of porosity and void ratio

$$e = \frac{f}{1-f};\ f = \frac{e}{1+e}$$

5. Mass wetness or gravimetric moisture content (θ_g)

Mass wetness is often referred to as gravimetric water content.

$$\theta_g = \frac{M_W}{M_S}$$

Where, M_s represents the mass of soil dried in an oven at 105°C for 24 hours or until it looses no more water. The ranges from 25-60%, varying with variation in bulk density of the soil.

6. Volume wetness Or Volumetric moisture content (θ_v)

The volume wetness often referred to as volumetric moisture content computed as the total volume of soil occupied by water.

$$\theta_V = \frac{V_W}{V_t} = \frac{V_W}{V_S + V_f}$$

In sandy soils, θ_v at saturation is 40 %, in medium textured soil it is 50% while in clayey soils it is 60 % (approximately).

The use of θ_v instead of θ_g is conveniently used. It is used for computing fluxes and water quantities added to the soil by irrigation or rain or to the quantity subtracted from the soil by evaporation, transpiration or drainage.

Relationship of mass wetness with volume wetness

$$\theta_V = \theta_g \frac{\rho_b}{\rho_b}$$

Where ρ_w is the density of water, which is 1.0 Mg m^{-3} at 4°C and at atmospheric pressure. In mineral soils, bulk density is greater than density of water, so volume wetness exceeds the mass wetness.

$$\theta_V = \theta_g \rho_w$$

Relationship between air filled porosity and volume wetness

$$f_a = \frac{V_a}{V_t} = \frac{V_f - V_w}{V_t} = \frac{V_f}{V_t} - \frac{V_w}{V_t} = f - \theta_V$$

7. Degree of saturation (θ_s)

It is the index to express the volume of water present in the soil relative to the volume of pores.

$$\theta_S = \frac{V_W}{V_F}$$

In dry soil, degree of saturation is almost zero and is around 100 % at saturation. The complete saturation (100%) is rarely attained as some air may even present as entrapped in fully saturated soils. The is not a good index for swelling & shrinkage type of soil, since the porosity changes with wetness.

Relationship of degree of saturation to volume wetness

$$\theta_S = \frac{V_W}{V_f}$$

$$\theta_S = \frac{V_W}{V_f} x \frac{V_t}{V_t}$$

$$\theta_S = \frac{\theta_V}{f} [as\ \theta_V = \frac{V_W}{V_t}\ ;\ \frac{V_t}{V_f} = \frac{1}{f}]$$

8. Depth of water (d_w)

$$d_w = \frac{V_w}{A}$$

where A is the area (cm^2)

As, $V_t = Ad_t$

$A = \frac{V_t}{d_t}$; where d_t is the depth of total soil

$$d_w = \frac{V_w}{V_t/d_t} = (\frac{V_w}{V_t})d_t$$

$$\text{or, } d_w = \theta_v\, d_t$$

$$d_w = \theta_g\, \rho_b d_t$$

or,

,

(θ_g - mass wetness or gravimetric water content; ρ_b = bulk density of the soil)

Chapter 8

Soil Texture

Soil consists of solid, water and air. The solid phase of the soil is fixed in its composition and does not change with time but content of water and air is highly dynamic and changes due to reactions taking place. The solid component of the soil is made up of crystalline and amorphous mineral matter varying in size of varying degree. The amorphous compound which is attached to the mineral particles and sometimes coating the particles is generally small, therefore, the solid phase is considered as consisting of distinct mineral particles. The biggest mineral particles (sand and silt fraction) can be seen through the naked eye and the smallest (colloidal fraction) can be seen through electron microscope.

Solid phase forms the matrix and responsible for various properties of soil, therefore, the major interest is in finding shape, size & chemical/mineral composition of individual particle.

Soil separates: The mineral particles of a soil do not change readily. They are divided into three components based on their size i.e. sand, silt and clay. So, soil separates may be defined as the individual size group of mineral soil particles (sand, silt and clay) which are 2.0 mm or less in diameter.

Soil texture: The term soil texture represents the prominent size group of particles which can be defined qualitatively as well as quantitatively as:

"*Qualitatively* soil texture is defined as the feel of soil when moist whether it is coarse or fine when rubbed between thumb and fingers".

"*Quantitatively*, soil texture refers to the relative proportions of different size groups of soil separates, i.e. relative proportion of sand, silt and clay in soil".

Since soil texture cannot be changed easily therefore it is considered as a static/fixed property. Hence, texture referred to as a basic/inherent property of a soil influencing all the soil properties. It determines the economic value of the land to an appreciable extent. Capability of the land for its use and methods of soil managements are largely dependent on soil texture.

Mechanical composition of soil: The mechanical composition of the soils refers to the estimation of different fractions of soils i.e. basic components viz. sand, silt and clay but not aggregates and organic particles.

Classification of soil particles: Earlier the classification of soil particle was based on shape, density, and chemical composition of the particle but is obsoleted and not in use in present time. Now the classification of soil particles is entirely based on its size. The most important criterion of determining texture of the soil is the size range limit of the different particle that is to be added to the definition of soil material. Some soils contain large rocks that obviously do not behave like soil but if large rocks are numerous then they might affect the behavior of the soil in bulk. The *soil material* are considered as the particles < 2.0 mm in diameter. This material is also called *fine earth*.

The fine earth particles are considered in mechanical analysis of soils. Different agencies have classified the soil texture based on the size of the soil separates, of which classification given by "International Union of Soil Science (ISSS)" (earlier known as International Society of Soil Sciences) and the "United States Department of Agriculture (USDA)" has been widely adopted or in use.

Classification of soil separates based on its size

International Union of Soil Science (IUSS) System– Atterberg 1912			
Sand (mm)		**Silt (mm)**	**Clay (mm)**
Coarse	**Fine**		
2.0 - 0.2	0.2 - 0.02	0.02 - 0.002	< 0.002

United States Department of Agriculture (USDA) System						
Sand (mm)					**Silt (mm)**	**Clay (mm)**
V. Coarse	**Coarse**	**Medium**	**Fine**	**Very fine**		
2.0 - 1.0	1.0 - 0.5	0.5 - 0.25	0.25 - 0.1	0.1 - 0.05	***0.05 - 0.002***	< 0.002

British standard Institute (BSI) System						
Sand (mm)			**Silt (mm)**			**Clay (mm)**
Coarse	**Medium**	**Fine**	**Coarse**	**Medium**	**Fine**	
2.0 - 0.6	0.6 - 0.2	0.2 - 0.06	0.06 - 0.02	0.02 - 0.006	0.006 - 0.002	< 0.002

The IUSS classification was first suggested by Atterberg (1912). In these classifications, upper size limit of clay fraction of 0.002 mm is based primarily on the expression of the physico-chemical properties associated with surface activities of particles. Clay is an active fraction of the soil which may possess definite chemical and physical activities. On the other hand, sand and silt separates don't exhibit marked physical and chemical activities. The relatively inert sand and silt fractions may be called *the soil skeleton* while the clay, by analogy, can be regarded as *the flesh* of the soil. All three fractions of the solid phase constitute *the matrix* of the soil as they are combined in various configurations.

Particles with size > 2.0 mm are referred to as *gravel*, pebbles, cobbles or stones and boulders depending on their sizes as follows:

Gravels or pebbles	–	2-75 mm diameter
Cobbles	–	75-250 mm
Stones	–	250–600 mm
Boulders	>	600 mm

PHYSICAL PROPERTIES OF SOIL SEPARATES

1. **Size and shape:** The size limits of different soil separates have already been discussed in the preceding section. Sand and silt particles are primary minerals, generally spherical or cubical in shape. Unless coated with clay particles, the sand particles are not sticky when wet and cannot be moulded into desired shape and size. Sand particles have low water holding capacity due to dominance of macropores. Silt particles are intermediate between sand and clay with irregular shape. They may possess some plasticity, cohesion and adsorptive capacity. Clay particles are platy or needle like and are secondary minerals. The surface area per unit mass of clay particles is very high due to small size of the individual particles. They may possess considerable stickiness and plasticity properties when wet. They may expand and contract on wetting and drying and has high water holding capacity.

2. **Density:** Density of soil separates ranges between 2.6-2.7 Mg m^{-3} as these separates are composed of minerals like quartz, mica, feldspars, etc. having the density of about 2.65 Mg m^{-3}. Therefore, the average value of particle density of soils is nearly 2.65 g cm^{-3}.The particle density of soil may be higher if it contains large amount of minerals like iron oxide and other heavy minerals. Depending on the amount of humus, the particle density of soil may change as the density of soil organic matter is generally low (1.1-1.3 Mg m^{-3}).

3. **Mineralogical and chemical composition:** Silicate ions constitute the greatest part of the mineral fraction of soil. Sand and silt contain mainly primary minerals (predominantly quartz, SiO_2) which are highest in silica. Also, the other primary minerals which are usually found are feldspar & micas. Gibbsite, hematite and goethite are also found usually as a coating on sand particles. Silts consist particularly of K-bearing minerals. The clay is highly active and contains substantial amount of K, Ca and Mg, in addition to fe and Al oxides and hydroxides etc.

   The quartz has been the dominant minerals found in sand; whereas fine sand and silt are dominated by quartz and feldspars; and in clays minerals like vermiculite, montmorillonite, kaolinite and amorphous colloids dominates.

4. **Specific surface:** The specific surface is expressed as the ratio of surface area of solid soil particles to its mass or volume. Its units are cm^2/g or cm^2/cm^3 of soil particles.

   All the reactions physical and chemical in the soils viz. sorption of ions, water molecules, swell and shrink, plasticity, CEC, movement of nutrient ions in the soil and availability etc. depends upon the specific surface of the soil, the extent of these reactions and processes is proportional to the soil's specific surface.

For spherical particles:

$$Ss = \left(\frac{\text{Surfacearea}}{\text{Volume}}\right) = \left(\frac{4\pi r^2}{\frac{4}{3}\pi r^3}\right) = \left(\frac{3}{r}\right)$$

Where Ss is the specific surface area of a spherical particle (cm^2/cm^3) and r is the radius of the spherical particle (cm). If the particles are cubical in shape, then specific surface is:

$$Sc = \left(\frac{\text{Surfacearea}}{\text{Volume}}\right) = \left(\frac{6L^2}{L^3}\right) = \left(\frac{6}{L}\right)$$

Where Sc is the specific surface of a cubical particle (cm^2/cm^3) and L is the side of the cubical particle (cm).

With the decrease in particle size, the specific surface of the soil increases. For the same volume, the degree to which the size of particles decreases, an equal proportion of increase in the specific surface of soil results.

Let us takes an example of the particle with cubical in shape of length 1.0 and 0.1 cm, then

$$Sc = \left(\frac{6}{1}\right) = 6\ cm^2/cm^3$$

If we divide this cube into 10 smaller cubes whose sides are 0.1 cm or say 1 mm, then

$$\mathbf{Sc} = \left(\frac{6}{0.1}\right) = \mathbf{60}\ cm^2/cm^3$$

Therefore, the cubical shaped sand particle of size 1.0 mm would have a specific surface of 60 cm^2/cm^3 only, however, with the reduction in size of the particle from 1.0 mm to 0.001 mm as that of colloidal clays, the specific surface increased to 60,000 cm^2/cm^3 which is one thousand times (1000) times greater than specific surface of sand particles.

The plate like shape of clay particles would also increase the specific surface to some extent. Besides size and shape, type of clay minerals such as montmorillonite, kaolinite, illite etc. also affect specific surface of the soil.

Comparative specific surface of the different size particles of the soil

Size fraction/ soil	Diameter (mm)	Specific surface (cm^2/g)	Clay mineral	Specific surface (m^2/g)
Coarse sand	2.0-0.20	45	Kaolinite	37-45
Fine sand	0.2-0.02	446	Illite	120-170
Silt	0.0-0.002	4458	Chlorite	130-180
Clay	10^{-4} -10^{-6}	1000 x 10^4	Montmorillonite	580-750
Sandy loam		1x10^4- 4x10^4	Vermiculite	780-900
Loam		50x10^4- 100x10^4		
Clay		150x10^4-250x10^4		

5. **Other properties**: Sand particles exhibit practically very low plasticity, cohesion (stickiness), adsorption of water and nutrients, heat of wetting, surface tension, capillarity etc. On the other hand, clays exhibit high degree of these phenomena. Silt fraction possesses some plasticity, cohesion, etc., but to a much lesser degree than the clay.

EFFECT OF SOIL SEPARATES ON PROPERTIES AND BEHAVIOUR OF SOILS

The different soil separates have great effect on the properties of the soil and thus its behaviour. Soils may contain varying proportions of sand; silt and clay i.e. soil separates, and exhibit properties which depend upon its dominant fraction.

Influence of different soil separates on properties & behavior of soils

Property/behavior	Rating associated with soil separates		
	Sand	Silt	Clay
Feel	Gritty	smooth	Plastic
Water holding capacity	Low	Medium to high	High
Plant available water	Low	Medium	High
Aeration	Good	Medium	Poor
Drainage	High	Slow to medium	Very slow
OM decomposition	Rapid	Medium	Slow
Summer warming up	Rapid	Moderate	Slow
Compactability	Low	Medium	High
Wind erosion	Moderate (high in fine sand)	High	Low

Property/behavior	Rating associated with soil separates		
	Sand	Silt	Clay
Water erosion	Low (unless fine sand)	High	Low if aggregated
Shrinking-swelling	Very low	Low	Moderate to very high
Sealing	Poor	Poor	Good
Fertility	Poor	Medium	High
Buffering capacity	Low	Medium	High
Runoff	Low	Low-medium	Medium-high

DETERMINATION OF SOIL TEXTURAL CLASS

There are two methods to determine the soil textural class

1. **Feel method:** In field, soil texture may be determined by rubbing or feeling the moist soil between the thumb and forefinger. The common characteristics of soil taken into being consideration are coarseness (grittiness), cohesion (attraction of like molecules), adhesion (attraction of unlike molecules), plasticity (formation of ribbon) etc.

Soil with high content of sand feel rough and gritty and makes a grinding noise when rubbed near ear. Soil with high content of silt makes it smooth and silky and looks like flour or talc when dry with very little stickiness when wet. A soil with very high content of clay will feel sticky and that forms ribbon when pressed between thumb and fingers.

The feel method is used by soil survey scientists but it need an expertise hand for its determination.

Steps to be followed in determining the texture of the soil

- ◈ **Type of ribbon** (poor, medium or good) – Clay, Clay loam, Loam, sandy loam class etc.

  No ribbon – Sand, Loamy sand

- ◈ **Grittiness or smoothness**
  - Grittness (Hand) – dominance of sand – Sandy Loam, sandy clay loam or Sandy clay
  - Smoothness - Silt loam, Silty clay loam or Silty clay
  - Neither gritty nor smooth – Loam, Clay loam or Clay

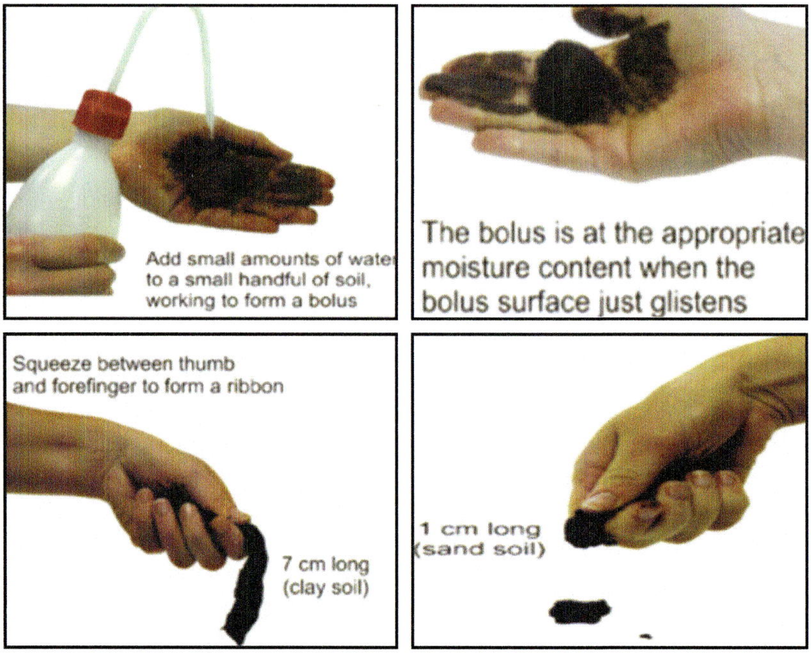

(source: https://www.agric.wa.gov.au/soil-constraints/soil-texture-estimating-hand)

2. **Laboratory method:** The U.S. Department of Agriculture has developed the method for determining particle size analysis for naming the textural class of the soil. It involves two steps
   a) Mechanical analysis
   b) Determination of soil textural class using equilateral triangle

a) Mechanical Analysis

The process of determining the various amount of different inorganic mineral soil separates i.e. sand, silt and clay in a soil by sieving, sedimentation or combination of both is called as mechanical analysis.

This procedure involves two major steps:

1. Separations of particles from each other i.e. complete dispersion into ultimate particles.
2. Fractionation is estimating the proportion of each sized fraction of the soil sample.

1. **Separation:** Preparation of samples for complete dispersion of aggregates in to their individual primary particles without breaking the particles. It is done by removing the soil binding /cementing materials viz. organic matter, $CaCO_3$, and/or oxides of Fe & Al, and subsequently deflocculation of the clay particles.

a) ***Removal of binding materials***: The most important binding agent in the soil is organic matter which can be removed by oxidizing the organic material with hydrogen peroxide (H_2O_2). Other binding agents are $CaCO_3$ and oxides of Fe & Al which are removed by treating the soil with dilute HCl. The following reactions occur and the binding agents are converted into gas or brought into soil solution.

$$O.M. + 2H_2O_2 = CO_2 + 2H_2O$$

$$CaCO_3 + 2HCl = CaCl_2 + CO_2 + H_2O$$

$$Al_2O_3 + 6HCl = 2AlCl_3 + 3H_2O$$

b) ***Removal of salts:*** Salts which are already present in the soil and those formed during reactions with dilute HCl are removed through filtration with distilled water. The silver nitrate ($AgNO_3$) test is usually performed to ensure complete removal of chlorides from the sample. The $AgNO_3$ gives white precipitation in the presence of chloride in the solution.

c) ***Separation of clay particles:*** Mechanical stirring in water suspension separates the clay particles from each other.

d) ***Deflocculation of clay particles:*** Once the clay particles are separated, they should remain separated but they have the tendency to re-unite. Therefore, to keep the clay particles deflocculated, small amount of a dispersing agent such as NaOH or sodium hexa-meta-phosphate or with Calgon solution (Calgon solution consists of sodium hexa-metaphosphate with sufficient amount of sodium carbonate to give pH > 8.3) is added.

2. **Fractionation:** Fractionation means separation of soil particles into different size groups and measurement of the amount of soil material in each size group. Different size sieves are used for separating out very fine sand and larger separates from finer silt and clay particles. Then the weight of each separate of different size is measured. Usually, the relation is used to convert mesh number into size of sieve openings or the particle size in mm.

   For separating still smaller particles (<0.02 mm), the method of sedimentation is usually used which is based on measuring the relative settling velocity of particles of different sizes in an aqueous suspension. Sedimentation is the settling of particles in a fluid under the influence of gravity or centrifugation.

The important sedimentation methods are

a) **International Pipette method:** The suspension of soil is stirred, made to stand on a smooth surface and at varying time intervals, sampling is done through pipetting. The different particles of differing size fall through the fluid at different rates and time of fall is calculated by Stoke's Law.

The measured quantity of soil suspension is pipetted out from the fixed depth below the surface of suspension and kept in a container for drying it in an oven at 105 °C for 24-48 hours or upto the time when water evaporates completely and weighed. This way the amount of silt + clay and amount of clay can be estimated.

So. "*International pipette method* is based on the principle of sedimentation i.e. different size particles having different weights fall with different velocities".

Advantages and disadvantages of pipette method

This method of soil textural analysis is the accurate methods of estimating sand, silt and clay, so is considered as the standard method.

It consumes lot of time in its analysis; therefore, this method is not suitable for its textural analysis when the bulk of samples are to be analyzed.

b) ***Hydrometer method*** (Bouyoucos, 1927): The dispersed (treating the soil with NaOH) soil sample after sieving is transferred to a sedimentation cylinder and filled upto the mark. After stirring, add two drops of Octan-2Ol (alcohol) to control frothing caused by agitation. Hydrometer is immersed into the suspension and the readings are noted at varying intervals of time for estimating silt + clay and clay fractions of the soil.

Thus, "it is based on principle of that there is continuous decrease in density of the soil suspension with time at the rate the particles fall below the level of hydrometer".

Advantages and disadvantages of hydrometer method

The hydrometer method of soil textural analysis is faster and moderately accurate and is adopted when large numbers of samples are to be analyzed. This method is not accurately used for saline, organic and calcareous soils or soils which are difficult to disperse as the pretreatment of soil with H_2O_2 and dilute HCl is not to be done and there is no filtration process is adopted for the removal of salts.

Stokes' Law (*G.G. Stokes, 1851*): "The terminal velocity or settling velocity of a spherical particle falling under the influence of gravity in a viscous fluid of a given density and viscosity is directly proportional to the square of the effective radius of the particle, gravitational force (g), difference between the density of the particle and density of the fluid (and inversely proportional to viscosity".

$$V = \frac{2}{9} gr^2 \frac{(\rho_s - \rho_f)}{\eta} \qquad (1)$$

Where

V = terminal velocity of particles falling through the viscous fluid, cm/sec

ρ_s = density of the particle, Mg m^{-3}

ρ_f = Viscous fluid density, Mg m^{-3}

g = acceleration due to gravity, cm sec^{-2}

r = radius of the falling particles, cm

ç = viscosity (poise)

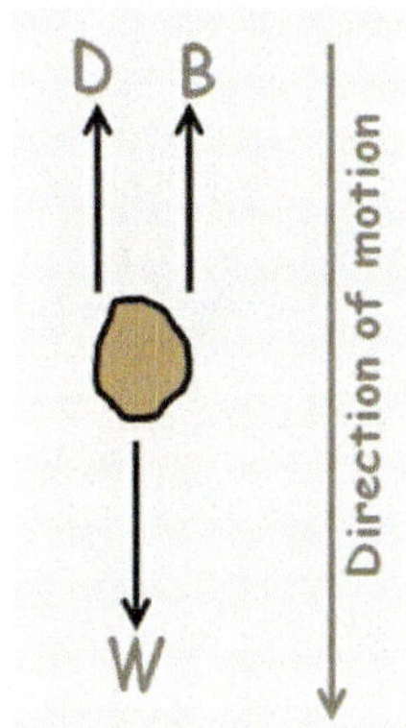

Source: https://www.geeksforgeeks.org /stokes-law/

Derivation of Stokes' Law:

In a vacuum, the particle (spherical in shape) falling will not experience any resistance force and its velocity will be accelerated through gravity only. Thus, the velocity will increase during its fall in the vacuum On the other hand; the particle falling in a viscous fluid will come across a frictional resistance (F_r) force which acts in direction (upward) opposite to the direction of fall of the particles. This fall of the particle in the viscous fluid is directly proportional to its radius (r) and velocity (v) of the particle and viscosity of the fluid (η). Thus the frictional resistive force is given as

$$F_r = 6\pi \eta rv \qquad (2)$$

At first, when the particle starts to fall in the fluid, velocity of fall increases due to gravity and in a due course of time, a point will reach at which the frictional resistance force (*drag and buoyant force*) acting in the upward direction to the fall of the particle becomes equal to the downward force due to gravity.

At this stage, the acceleration becomes zero and the particle starts falling with a uniform velocity called as the *terminal velocity or settling velocity.*

The gravitational force (F_g) acting downwards on the particle is the product of its mass (m, gm) and acceleration due to gravity (g, cm/sec^2)

$$F_g = m \times g$$

$$F_g = \frac{4}{3}\pi r^3 \rho_s . g$$

[as volume of sphere = $\frac{4}{3}\pi r^3$ and mass = density x volume]

The buoyant force (B) (F_b), due to displacement of fluid by the particle also acts in the upward direction resisting the motion of the particle, is the weight (W, gm) of the displaced fluid

$$F_b = W = \frac{4}{3}\pi\, r^3 \rho_f . g$$

[as W = force = mass x acceleration due to gravity]

Where ρ_f is the density of water/fluid

As, $\rho_s > \rho_f$, F_g is more than F_b and the soil particles will fall downwards. So the resultant downward force acting on the soil particles is

$$F_d = (F_g - F_b) = \frac{4}{3}\pi r^3 \rho_s . g - \frac{4}{3}\pi\, r^3 \rho_f . g$$

$$F_d = \frac{4}{3}\pi\, r^3 (\rho_s - \rho_f) g$$

When a particle starts to fall, the velocity of fall increases up to a certain point. At equilibrium, the driving force (F_r) equals to the downward force (F_d) and the particles continues to fall without any acceleration i.e. at constant velocity.

$$\text{So, } F_d = F_r$$

$$6\pi \eta r v = \frac{4}{3}\pi\, r^3 g\, (\rho_s - \rho_f)$$

$$\text{Or } V = \frac{2}{9}\, g r^2 \frac{(\rho_s - \rho_f)}{\eta} \qquad (4)$$

This equation describes the Strokes law

At specific temperature for a particular fluid and falling particles, ρ_s, ρ_f, η and g are kept as constant. Thus, the equation (4) can be written in a simple form as

$V = kr^2$, where k is a constant (5)

Thus, the sedimentation velocity of the soil particle is directly proportional to the square of the radius of that particle.

Applicability or use of Strokes law

The Stokes' law is applicable to a solid spherical shape soil particle falling through a viscous liquid or gas or to a drop of liquid falling through a gaseous medium.

The time taken for the particle to fall from the certain height (h) in the viscous fluid can be calculated from the Stokes' law as :

$$t = \left(\frac{18\, h\, \eta}{d^2 g\, (\rho_s - \rho_f)} \right)$$

where d is the diameter of the particle

$$\text{as } v = \frac{h}{t} \text{ and } r^2 = \frac{d^2}{4}$$

Rearranging the equation (6):

$$d = \sqrt{\frac{18\, h\, \eta}{t\, g\, (\rho_s - \rho_f)}} \qquad (7)$$

Particle size distribution can be measured by pipetting out a known volume of soil suspension from the fixed depth at specific interval of time period as the sedimentation starts, the method commonly known as **International Pipette Method** of soil textural analysis. The other method is to use a **hydrometer** to measure the density of the suspension at a given depth as function of time. With time this density decreases as the largest particles and then progressively smaller ones, settle out of the region of the suspension being measured.

Assumptions and limitations of Stokes' Law:

1. *Particles of the soil should be spherical shaped, smooth (plane) and firm.* But, clay minerals are plate like shaped and the rate of its fall is comparatively slower than the same mass of the spherical particles.

2. *The particles of the soil should be larger than the size of the liquid molecules, so* that Zig-zag, chaotic haphazard movement i.e. Brownian movement will not affect the fall of the particles. The particles of the size 0.0002 mm or less will experience the Brownian movement. Particles ranging from 0.10 to 0.0002 mm in size can be estimated using sedimentation method. Soil particles less than 0.0002 mm can be separated using a centrifuge.

3. *Fall should not be hindered due to other smaller particle or due to walls of the container.* The large particles fall faster, may push the finer particles downwards along with them, if the soil suspension ratio exceeds > 3.0 % by volume. Also, particles falling along the walls of the container (0.1 mm distance) are also slowed down due to proximity of the walls of the container. Thus, the soil suspension ratio of < 3% of solids may lesson such hindrances.

4. *Density of the particles must be uniform.* Most of the soils have uniform density of 2.65 Mg m^{-3} due to existence of minerals like quartz, feldspar etc. but it may increase due to existence of the Iron and aluminium oxides.

5. *The particles must be sufficiently small for the suspension to create any turbulence.* Particles of the size >0.05 or 0.02 mm diameter settle quickly due to which it can cause turbulence, thus can be removed through sieving procedure.

SOIL TEXTURAL CLASSES

Textural class refers to the overall designation of the texture of a soil. Soil textural class is determined on the relative proportion of sand, silt and clay on weight basis. Soils with varying proportions of the soil separates i.e. sand, silt and clay have been assigned different classes as designated in *Textural triangle* (Fig. below).

The three *broad primary textural groups* of soils within which all the classes lies are sandy soils, loamy soils and clayey soils. Within each of these three broad groups, specific textural classes have been categorized based on the relative proportional of sand, silt and clay particles. These textural classes are divided into 12 categories.

Demonstration of the equilateral textural triangle

Textural triangle can be demonstrated by taking an example of the soil that contains 45% sand, 40% silt and 15% clay. To the per cent clay side of the textural triangle, point out 15% clay and draw a line from this point parallel to the per cent sand side of the trainagle i.e. base of the textural triangle. Now on the per cent sand side of the triangle i.e. at the base of the triangle, point out 45% sand and draw a line parallel to per cent silt side of the triangle. The two lines intersect at a point corresponding to 40% silt. The different compartments representing the soil textural classes in which these lines intersects indicates the textural class in which that soil sample lies i.e. in this case it is *loam*.

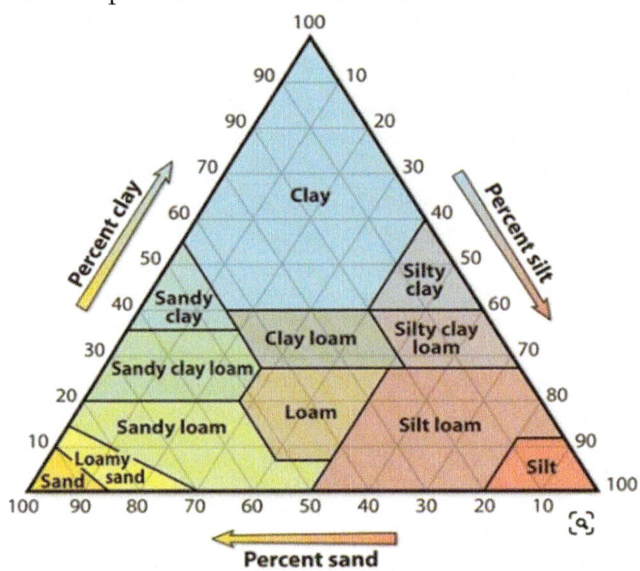

Fig. Textural triangle (sand, silt and clay particle sizes of 2.0-0.02 mm, 0.02-0.002 mm and < 0.002 mm, respectively)

(Source: https://www.kisaanhelpline.com/research-article/165/STUDY-AND-ANALYSIS-OF-SOIL-TEXTURE)

Broad primary textural group to describe texture in relation to basic soil textural class names

Textural group	Texture	Soil textural class
Sandy soils	Coarse	sand loamy sand
	Moderately coarse	sandy loam fine* sandy loam
Loamy soils	Medium	very fine sandy loam* loam silt loam silt
	Moderately fine	sandy clay loam clay loam silty clay loam
Clayey soils	Fine	sandy clay silty clay clay

* *not included in textural triangle diagram. These soils are usually treated separately because of their fine sand content*

1. ***Sandy soils:*** Sandy soils are essentially those soils in which and content is 70% or more with less than 15 % clay content of the material by weight. Two specific textural classes included in the sandy soils are

    ii) *sand*

    iii) *loamy sand*

Sandy soils are single grained structure i.e. loose, low water retentive, well aerated and properly drained and easy to cultivate. These soils are also called as *light* soils.

The expression of light and heavy soil is based on the relative energy required to operate tillage implements for preparation of seed-bed tilth and other soil operations. Energy required for tillage is much less in light soils as compared to that in heavy soils. These terms are no longer in use now as they are little confusing. The coarse textured soils are generally denser than the fine textured soils, and thus, are heavier in weight per unit volume as compared to fine textured soils.

In general, sandy soils because of dominance of macro pores, have low water and thus nutrient holding capability, low organic matter content, almost nil swelling and shrinkage type of clays, poor sealing properties for ponds and dams, high leaching of nutrients and pollutants. The erosion of fine sands is easily carried out by wind while coarse sands even resist erosion by water.

Loamy soils: In this group, following seven soil textural classes exists

i) *sandy loam,*

ii) *loam,*

iii) *silt loam,*

iv) *silt,*

v) *sandy clay loam,*

vi) *clay loam* and

vii) *silty clay loam.*

To qualify for the modifier sandy or silt, a soil must have atleast 40-50% of these separates. Thus, a loam in which sand is dominant is classified as sandy loam.

"An ideal loam is defined as a mixture of sand, silt and a clay particle which exhibits the properties of these soils separates in about equal proportions". This does not mean that the three separate are present in equal proportion, rather the three fractions of the soil exhibits properties which must have equal impact.

Loam soils is considered important for agriculture as these soils have more water holding capacity than sandy soils thus has the higher capacity to make nutrients available to the plants while its drainage and aeration status is better than that of clayey soils. Also, these soils are easy to till than that of clay soils which can make the tilth of the soil for proper and optimum growth of the plants. However, sandy and clayey soils may become more suitable than loam soils under certain specific environmental conditions and with proper management practices.

If these soils contains dominating silt content, then they may exhibits better soil physico-chemical properties viz. considerable capacity to hold water and thus have higher nutrient supplying capacity as compared to sandy soils, better aeration status, slow to medium drainage, possess medium organic matter content, capability to leach pollutants and nutrients. However, these soils are susceptible to erosion through wind and water, can be easily compacted, have little capability to swell and shrink and needs a power to till when wet or after rain.

Clayey soils: A soil which essentially contains at least 35% of the clay separates are classified under clayey soils. The textural class in this group includes

i) *sandy clay i.e.* per cent of clay is 40% or more,

ii) *silty clay* or

iii) *clay*

Sandy clay textural class contains high proportion of sand than that of clay. Similarly, silt content of silty clay textural class has its higher proportion than that of clay fraction.

The properties of clayey/ fine textured soils are that they tend to absorb and retain water much higher due to the dominance of micro porosity. Also, these soils show plasticity and sticky phenomenon when wet at varying degree, but becomes very hard and cohesive on drying. These soils are difficult to cultivate so are also called *heavy* soils as huge energy is expended for the tillage implements to be operated.

The clayey soils have high capability to retain water and thus have higher capability to supply nutrients to the soils, poorly aerated; have very poor drainage due to more microporosity, form deep wide crack under dry conditions, higher amount of organic matter content, medium to high swelling and shrinkage. These soils resist wind erosion and aggregated clays also resist water erosion. They have very good sealing properties. Theses soils can be compacted to higher degree and translocation of nutrients and pollutants is very slow due to which groundwater pollution through soils is very less. Under extremely wet and dry conditions, these soils are very difficult to till.

Chapter 9

Soil Structure

The *primary particles* (sand, silt and clay) do not found as an individual entity in nature but exist as larger units or aggregates through joining together usually called as *secondary particles*. These aggregates are defined as the naturally occurring semi-permanent groups of soil particles. The binding force between intra-particles in an aggregate is much stronger than the forces between adjacent aggregates.

Peds: "Naturally formed soil aggregates formed by natural force of attraction between the particles are also called *peds*"

Clod: "Artificially formed irregular shaped coherent mass of soil formed due to tillage operations is called as *clod*".

Fragment: "When a ped break along the natural plane of weakness, then each piece of broken ped is called as *fragment*".

Soil structure: "*Soil structure may be defined as the arrangements of primary and secondary soil particles into some structural pattern or shape"*. The arrangement of these soil particles lead to the development of pores of varying sizes. Therefore, the soil structure may also be defined as the arrangement of various sized soil pores into some structural pattern.

Edaphologically, Soil structure refers to the secondary units called aggregates composed of primary soil particles which control the pore size distribution, stable when wet and hardness when dry.

Significance/importance of soil structure: Soil structure influences all the factors of plant growth inspite of the fact that it does not have direct impact on plant growth factor for eg: availability of water, exchange of gases from soil to atmosphere and vice versa, movement of plant nutrients, heat conduction, penetration of roots, microbial activity, etc. Soil management practices such as tillage practices, cultivation of crops, fertilization &, manuring, amendments (liming, gypsum) and irrigation are also influenced by soil structure.

Classification of soil structure: Many attempts have been made to classify soil structure but yet there is not a universally accepted system of its classification. It is because there is no satisfactory method of measuring soil structure. The methods available are in fact indirect methods i.e. measuring one or more soil attributes which are dependent upon soil structure.

Soil structure can be described and classified based on

i) *Type of structure* which is based on the shape & arrangement of the peds;

ii) *Class of the structure* which is based on the size of the peds; and

iii) *Grade of the structure* which is based on the distinctiveness and durability of the peds.

Soil structure may also be classified according to pore sizes.

TYPES OF SOIL STRUCTURE

A. Based on the shape and arrangement of peds

Soil structure is classified into simple and compound structures (International Committee of Soil Science).

1. **Simple structure:** Simple structured soils are those in which natural cleavage planes are indistinct or absent. Simple structure is of two types.

   a) ***Single grained structure:*** When the particles are completely unattached to each other or structure is completely loose, the soil is said to be structureless. Since even a loose arrangement is also a sort of structure, the term single grained structure is preferred. Single grained structure is found in loose sand of desert regions of the world.

   b) ***Massive structure:*** When the particles are bonded in large cohesive, non-structured blocks, it is called as massive structure. It is similar to single grained structure except it is coherent and particles are attached to each other e.g. surface crust, plough pan, clay pan.

2. **Compound structure:** In compound structure, natural cleavage planes are distinct or visible with naked eyes. The shape of individual ped is described according to the relative length of the vertical and horizontal axes, and by the shape of their edges. Compound structures are classified into four principal types.

   a) ***Cube or block like*:** When vertical and horizontal axis are of similar lengths and peds are developed equally along the three axis, the structure is considered as *block or cube like*. It is further sub-divided into two groups:

i) *Angular blocky*: Flat faces with sharp angular edges and corners.

(Source: https://wiki.ubc.ca/LFS:SoilWeb/Interactions_Among_Soil_Components/Soil_Structure)

ii) *Sub-angular blocky*: Faces and edges are mainly rounded.

The blocky structure is observed in sub-surface horizons (B-horizon) due to which it promotes good drainage of water, air movement and easy root growth in humid region.

b) ***Prism like***: The aggregates are elongated in vertical direction i.e. vertical axes are more developed than the horizontal axes, giving a column or prism like shape of the ped. Vertical cleavage planes predominant. It is again sub-divided into two groups.

i) *Prismatic*: When the tops of peds are relatively angular and flat.

ii) *Columnar*: When tops of peds are distinct and rounded.

Prism like structures is commonly found in sub-soil horizons in arid and semi-arid regions.

(Source: https://wiki.ubc.ca/LFS:SoilWeb/Interactions_Among_Soil_Components/Soil_Structure)

c) ***Plate-like***: In this, the axes along the horizontal axes are longer than that of vertical axes, resulting in a plate like peds. Horizontal cleavage planes predominate. When unit are thick, they are called *platy*, and when units are thin, they are called *laminar*. The platy structure is often inherited from the parent materials. This type of structure is found in compacted soils or soils with high clay content. It may also be found in surface layers of some virgin soils.

(Source: https://wiki.ubc.ca/LFS:SoilWeb /Interactions_Among_Soil_Components/ Soil_Structure)

d) ***Spheroidal*:** All axes are developed equally with the same length. The peds are rounded with curved and irregular faces. Generally, all rounded or sphere like aggregates ranged between 2.0 -10.0 mm in diameter. It is again sub-divided into two groups:

   i) *Granular*: It has rough surface and is less porous due to low organic matter.

   ii) *Crumb*: When granules are especially porous due to high organic matter.

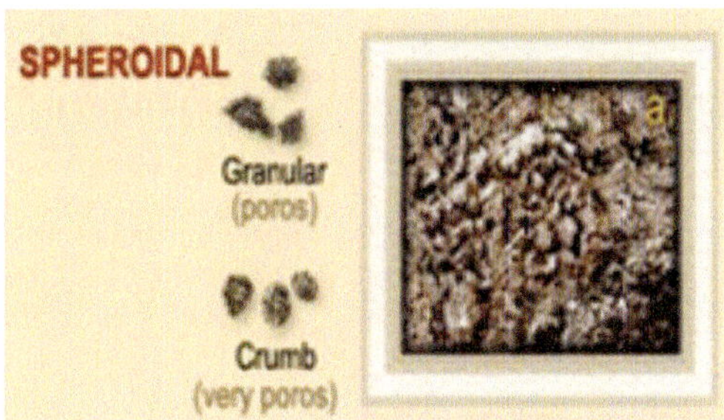

(Source: https://wiki.ubc.ca/LFS:SoilWeb /Interactions_Among_Soil_Components/ Soil_Structure)

The granular and crumb structure are characteristics of surface (A) horizon. In spherical structure, the physical properties like infiltration, percolation, aeration, etc., are not restricted.

Classes of soil structure: Each type of soil structure is differentiated further in the following classes on the basis of sizes of their aggregates (Nikiforoff, 1946).

Very coarse	>10 mm
Coarse	5-10 mm
Medium	2-5 mm
Fine	1-2 mm
Very fine	<1 mm

Classification of soil structure based on classes i.e. size of soil aggregates is a mean of quantitative classification. Aggregates of optimum size are favourable for plant growth than small or large size aggregates. A soil made up exclusively of silt size aggregates cannot be drained by gravity, since the pores are too small.

Grades of soil structure: Classification of soil structure based on grades is a mean of qualitative classification. The grades of soil structure are determined by the durability and stability of aggregates and the ease of separation by impact of rain-drops or under submerge condition. There are five grades: poorly, weakly, moderately developed, well developed and strongly developed structure.

The hardness of the aggregates is affected by moisture contents, amount & type of clay, nature of the adsorbed cations, and organic matter content of the soil.

For naming a soil structure, the sequence of grade, class and type is followed e.g. strong coarse angular blocky.

Classification of soil structure according to pore size

Since plant roots and microbes live in the pore spaces, therefore, the classification of soil structure according to pore size may be quite logical. The percentage distribution of pore space of various sizes is determined by removal of water from a saturated soil sample which gives a picture of structural pattern of a soil.

GENESIS OF SOIL STRUCTURE

Genesis of soil structure refers to *the causes and methods of formation of structural units or aggregates.* Two processes are involved in genesis of structure:

1. **Flocculation:** It's an electro-kinetic process involving both positive and negative charges. Primary particles with a high zeta-potential repel each other when they collide in a suspension. On lowering the zeta-potential sufficiently, a collision between particles results in mutual attraction and formation of floccules. Flocculating agents present in the soils makes the floccules stable. Flocculation is the grouping of the colloidal particles in a suspension. Colloidal particles of clay usually carry negative charges. When they come in contact with cations, floccules of silt size are formed. Flocculation also occurs due to dehydration, high soluble salt concentration in the diffuse double layer and presence of divalent or trivalent cations.

2. **Cementation:** "It refers to the consolidation/amalgamation of clay floccules by cementing material so that they are not dispersed once they are flocculated". The floccules are bonded with weak forces which can be easily broken so cannot make a stable soil structure unless they are further cemented by $CaCO_3$, oxides of Fe and Al, organic matter and humus in to the stable soil aggregates. For aggregation, flocculation of clay particles is very necessary but not a sufficient condition. ***Aggregation is flocculation followed by cementation.***

MECHANISMS OF AGGREGATE FORMATION

The mechanism of aggregate formation is highly complicated. The formation of stable aggregates requires the flocculation of colloidal material first and then cementation of floccules. Organic matter and clay particles when combined with appropriate cations on the exchange sites, the cations work as cementing material. Iron and aluminum hydroxides may also cement particles under some conditions. In calcareous soils, $CaCO_3$ binds the particles together. Mechanism of aggregate formation is described by the following theories.

1. **Russell's theory of crumb formation:** Russell (1934) proposed that clay particle in suspension is surrounded by a double layer of oppositely charges, the outer one being diffused and consisting of cations while the inner layer consists of negative charges on the surface of the particle. Around each negative charge on clay surface is an envelope of dipolar water molecules. Cations on the surface also possess envelopes of oriented water molecules. As the water is removed, the clay suspension becomes concentrated and the thickness of envelopes surrounding the negative charges on the surface of the clay particle and the cations is reduced. As a result each cation will share its envelopes with two clay particles and hold the particles together as shown in Fig. 7.1 below.

A linkage system is set up consisting of particle-oriented water molecules-cations-oriented water molecules-particle, known as *cation-dipole linkage.* As removal of water (dehydration) proceeds, an increasing proportion of cation share their water envelops with two clay particles and number of links increases. The links also become stronger because they become shorter and the cohesion of clay particles increases and crumbs are formed.

According to Russell's theory five conditions are necessary for crumb formation.

1. *High CEC*: Crumb formation is limited to particles possessing considerable CEC, indicating a need for high surface charge.
2. *Smaller particle size*: Crumb formation is limited to particles smaller than a certain size, sand and silt particles are not essential to crumb formation but tend to make the crumb mechanically weaker.
3. *Presence of dipole liquid in system*: Crumb formation is limited to those systems that contain liquid with an appreciable dipole moment e.g. water.
4. *Divalent cations*: The nature of exchangeable cations influences the hardness of the resulting crumbs. The relative order of effectiveness of the cations in reducing the hardness of the crumb is Mg>Ca>H>Na>K.
5. *Dehydration*: It is a basic requirement for the formation of crumbs

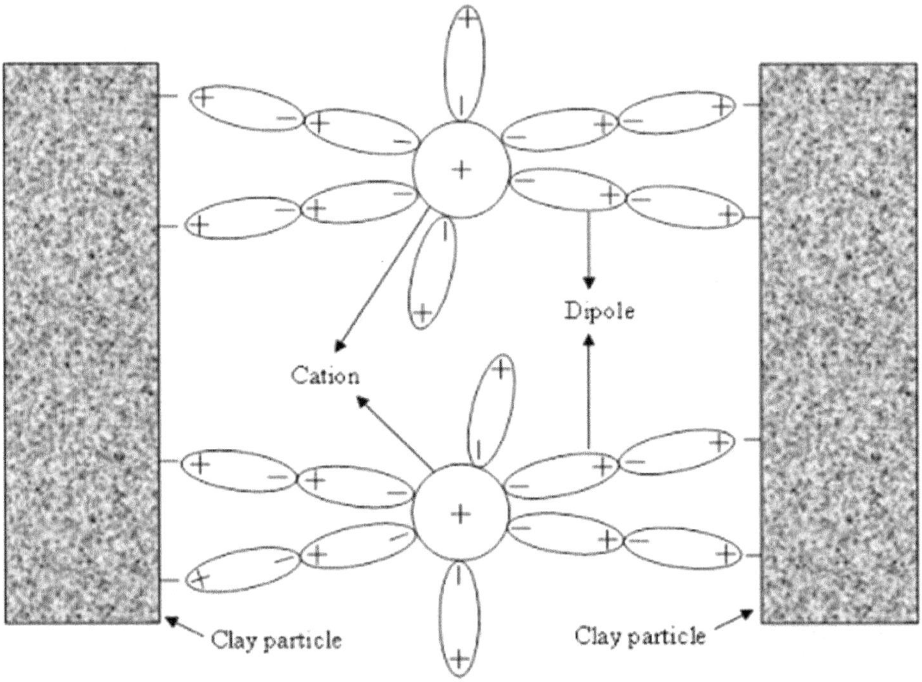

Fig. 7.1 Sketch of cation-dipole linkage between negatively charged colloidal particles

(Source: Text book of Soil Physics- A.K. Saha Edn)

2. **Emerson's theory of clay domain:** According to this theory, soil is considered to be made up of (i) clay domains representing the colloidal fraction and (ii) coarser particles of sand and silt that are represented by Quartz. Groups of clay particles form domains as a result of cation-dipole linkage (as described in Russell's theory). Emerson (1949) defined a clay domain as a group of clay particles having suitable exchangeable cations which are oriented and sufficiently close together by suitable cations for group to behave as a single unit in water.

The clay domains are bonded to the surface of the quartz as well as to each other to form crumbs. Colloidal organic matter forms complexes between the surfaces of the quartz-clay domain bonds, clay domain-clay domain bonds and quartz-quartz bonds to stabilize the crumb.

The generalized model of soil crumb is given in Fig. 7.2 in which various possibility of bonding arrangement in crumb formation are explained. The various types of bonds could be due to several kinds of bonding mechanisms:

1. Hydrogen bonds between carboxyl group of organic colloid and clay might occur. (Clay domain — H-OOC-R-COO-H — Clay domain)

2. Cationic Bonds:
   i) Between the carboxyl group of organic matter and the clay similar to hydrogen bonds given above.

      (Clay domain–Cation–OOC-R-COO–Cation–Clay domain)

   ii) Between carboxyl group of organic matter and quartz particles coated by colloidal particles.

      (Coated quartz–Cation–OOC-R-COO–Cation–Coated quartz)

   iii) Between carboxyl groups of organic matter, coated quartz and clay domains.

      (Coated quartz–Cation–OOC-R-COO–Cation– Clay domain)

   iv) Between clay domains. (Clay domain–Cation–Clay domain)

   v) Between coated quartz grains.(Coated quartz–Cation–Coated quartz)

3. Under flocculated conditions, attraction between particles due to van der Waals forces, may cause particles to unite.
4. A crumb once formed due to the bonding mechanisms may further be stabilized by irreversible dehydrated sesquioxides, if present in sufficient amount.

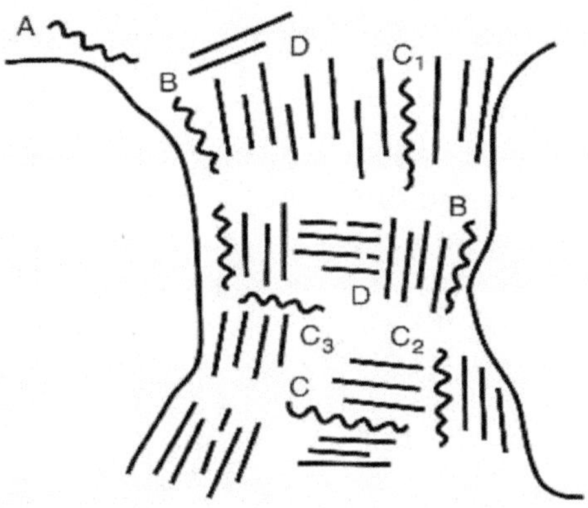

(Source: Text book of Soil Physics- A. K. Saha Edn)

Fig. 7.2 Generalized model of soil crumb (suggested arrangement of quartz particles, clay domains and organic matter in a soil aggregate). Types of bonds are: quartz–organic colloid–quartz (A); quartz–organic colloid–clay domain (B); clay domain–organic colloid–clay domain (C), face–face (C_1), edge–face (C_2), edge–edge (C_3); and clay domain edge–clay domain face (D).

It suggests that Emerson's model of soil crumb formation is an extension of Russell's earlier model as it describes the binding of various clay domains with each other and other components of the soil.

FACTORSINFLUENCINGTHEFORMATIONANDSTABILITYOFSOIL

1. **Amount and type of clay minerals:** Silicate clays assist in cementing primary particles into stable aggregates, therefore, higher the amount of clay more will be aggregation. Kaolinite clays flocculate and form aggregates more readily than montmorillonite. Montmorillonite clays form aggregates of cuboidal or spherical in shape whereas kaolinite clays produce platy structure

2. **Texture:** If a sandy soil contains a fair amount of silt and clay which is not sufficient enough for aggregation, thus may result in a massive structure upon puddling. Silt loam soils having a fair amount of organic matter may be well aggregated but with low organic matter these soils have unstable structure and crust easily. The soils of very high clay content also usually have massive structure due to high cohesiveness of clay particles. The massive structure is unfavourable for plants.

3. **Nature of exchangeable and soluble cations:** Calcium and magnesium are helpful in aggregation. Exchangeable Na in sodic soils deflocculates the soil colloids when wet. When dry, the soils are massive, hard and compact, and form crusts. If salt concentration is high, as in saline and saline-sodic soils, the soil colloids are flocculated. Such soils have very friable and good structure. But, due to high salinity, plant and microbial population is less and therefore, structure is not very stable.

4. **Organic Matter:** Organic matter plays a significant role for the formation of aggregates and making the soil structure more stable. Fats, waxes, lignin, proteins, resins, etc., have a direct effect in stabilizing soil aggregates. Organic matter, being loose and porous, reduces the cohesion and plasticity of clay, and is therefore, responsible for the porosity of soil aggregates. Humus serves as cement which form complexes with clays thus helping to form aggregates. The di and trivalent cation ions form complexes with organic compounds due to which floccules are stabilized. However, monovalent cation viz. K, Li and Na humus is deflocculated and does not help in soil aggregation.

5. **Dehydrated sesquioxides:** Hydrated sesquioxides (iron and aluminum oxides) form almost completely irreversible colloids upon dehydration and help to form stable aggregates.

6. **Other inorganic cement:** $CaCO_3$ precipitates on the soil particles and cement to form aggregation and imparts stability to soil aggregates.

7. **Synthetic soil conditioners:** Synthetic soil conditioners are long chain of carbon compounds can be used for soil aggregation. These long chain carbon compounds attach themselves to the exchangeable sites of the clay and link them together. Application of these synthetic soil compounds in small amounts can cement and stabilize aggregates.

These compounds are also used for reclaiming saline and sodic soils and producing good structure even in sandy soils.

8. **Plants and plant residues**: Plants can aggregate the soil particles in different ways. Plant roots excrete gelatinous compounds which bind the soil particles with each other. Root hairs also help the soil particles to stick together in granular form. Plant residues (roots and shoot) serve as food for microbes which play prime role in aggregate formation, and produce mucus substances on decomposition.

9. **Micro and macro-organisms:** The microorganism viz. bacteria, algae, fungi, actinomycetes, make the soil living through their activities. They help to bind the soil particles together by releasing gluey substances. They are more effective in binding the soil particles than that of exchangeable ion. These microbes also help in aggregation, indirectly, by producing cementing materials during decomposition of organic matter.

   The microorganisms like rodents, earthworms, spiders, mites, nematodes, insects, centipedes, millipedes, etc., help in the soil aggregation by burrowing, turning the soil or thoroughly mixing the organic residues with the soil.

10. **Water:** Water molecules acts as dipoles which help in orienting charged particles. It causes swelling and shrinking, and freezing and thawing of soil which help in developing planes of weakness and thus, formation of structure. Surface tension at air-water interface keeps the soil particles together. Sufficient water throughout the growing period assures maximum biological activity in the soil which is one of the prime requisites for the genesis of soil structure.

11. **Air:** Air affects aggregation through it physical, chemical and biological effects.

    ***Physical*:** Surface tension at water-air interface results in attraction of water with great force by dry aggregates. During wetting, the air is compressed in the centre of the aggregate and tends to escape with minor explosions. Unless large pores are present, aggregate breaks up.

    ***Chemical*:** Chemically, air assists in the precipitation of iron and aluminum collides that act as cementing agents. The high content of CO_2 in the soil air solubilize calcium and other ions in solution which form important links between soil particles.

    ***Biological*:** Biologically air is necessary for the respiration of roots and microbes. An extended period of flooding deteriorates soil structure.

12. **Temperature**: The type of clay formed and their zeta- potentials, exchangeable ions, soluble salts and organic matter are influenced by the temperature of the climate.

13. **Tillage:** Optimum tillage promotes aggregation. Tillage operations make the soil loose and mix the organic matter well with the mineral constituents. Excessive tillage using heavy machineries tend to destroy aggregation by disruption of weak aggregates, quick decomposition of organic matter and through compaction. Ploughing a heavy soil in wet condition destroys the soil structure and creates the puddled condition. When soils are cultivated in too dry condition, large clods in heavy and fine dust in light soils are produced.
14. **Irrigation:** The amount, method and time of irrigation also influence soil aggregation. Heavy irrigation breaks the aggregates. The broken aggregates block the soil pores and forms hard surface crust on drying. The slower rate of water application, as in case of sprinkler and drip irrigation systems, allow the aggregates to wet slowly and the explosive action of entrapped air becomes very small. While in fast rate of water application, as in case of flood irrigation, the aggregates are completely immersed in water instantly and the explosive action of air is very high.

When irrigation is applied in relatively moist soil condition, the dispersion of aggregates is very less. Sometimes irrigation water, unlike rainwater, carrying silt particles block the spaces between aggregates and forms the massive structure. In very sandy soils, these suspended materials may add sufficient silt to sand to allow aggregation.

EVALUATION OF SOIL STRUCTURE

The methods of evaluation of soil structure are grouped under two categories

1. **Direct methods** – It is based on measurement of shape, size and arrangements of soil aggregates. It is further divided into

i) **Microscopic method-** In this method, a thin section of soils are examined in a laboratory in petrography microscopic under polarized light which clearly shows the shape and size of secondary particles and its arrangements.

ii) **Macroscopic method (Field method):** In this method, a chunk of dry soil is extracted from the field and is made to fall on smooth surface at about 3 feet height. The chunk breaks in to small pieces of peds of different sizes and shapes. The geometric shape of the aggregates determines the type of structure. When the aggregates do not shows any definite line of cleavage, then the soil is called as structureless soils. Eg. Soils of sand dunes or along river or coastal belts where the particles are non-cohesive and single grained. Also, the structureless soils are formed in puddled soils of rice fields or in pans having dense and massive structure where the soil particles are highly cohesive and chunk do not break into small pieces.

2. **Indirect methods**-In these methods, soil structure is evaluated by determining the
   a) Size distribution and stability of aggregates commonly called as aggregate analysis
   b) Properties related to soil structure

Aggregate analysis

Aggregates analysis is carried out to measure the percentage of water stable secondary particles in the soil and the extent to which the finer mechanical separates are aggregated into coarser fractions. The techniques employed for aggregate analysis are:

a) **Wet sieving** (*Yodder type*) - In this method, soil is slowly wetted by capillarity for 30 minutes. The wetted soil sample is transferred in to the sieve having largest opening in the nest and the nest of sieves are immersed in water. The sieves are slowly raised and lowered vertically in the water by mechanical method for 30 minutes @ 30 oscillations / minute. The sieves in the nest have an opening from 8.0 mm to 0.1 mm.

   Oven dry weight of the fractions of the aggregates retained on each sieve is determined for knowing the proportionate distribution of the aggregates.

Yodder type wet siever

(Source: https://yatherm.com)

Precautions: Wetting of the soils should be done very carefully so that there may not be the breaking up of aggregates. Wetting by capillarity should generally be preferred but capillary action may also break the aggregates due to shattering effect of entrapped air. It is even better by wetting the soil under vacuum with de-aired water, which precludes the entrapped entrapment of air bubbles entirely. Spraying water in to the aggregates with an automizer produces less destruction of aggregates.

b) **Dry sieving** (*Chepil 1962*) – Dry sieving measures the size and stability of dry aggregates or clods which are weakly held together as in arid soils. The different size fractions of the soil passed through the series of sieves gives the aggregates size distribution.

Dry sieving apparatus

(*Source: https://yatherm.com*)

The principle of this method lies in determining the proportion of aggregates which are stable against vibrating action stimulating the scouring effect of wind.

Dry sieving of aggregates gives an important index for determining the susceptibility of soils to wind erosion.

c) **Elutriation method**: This method separates the different fractions of soil by washing out soil particles or aggregates in a rising current of water i.e. separating the lighter particles from heavier ones vertically oriented stream of liquid. Water is allowed to flow through vertical vessels with a given velocity in each successive vessels. The velocity of water flow is kept constant in all the vessels but velocity decreases when liquid water moves from one vessel to another due to increase in cross-sectional area of the vessel. Liquid water flows in the direction opposite to the sedimentation movement of particles. The particles having settling velocity less than the upward velocity of water in a particular vessel will be carried out of that vessel i.e. coarser particles will be retained in first vessel whereas finer particles in another two vessels. This method may be used for separating aggregates having diameter between 1.0 to 0.02 mm and is used where wet sieving cannot be adopted.

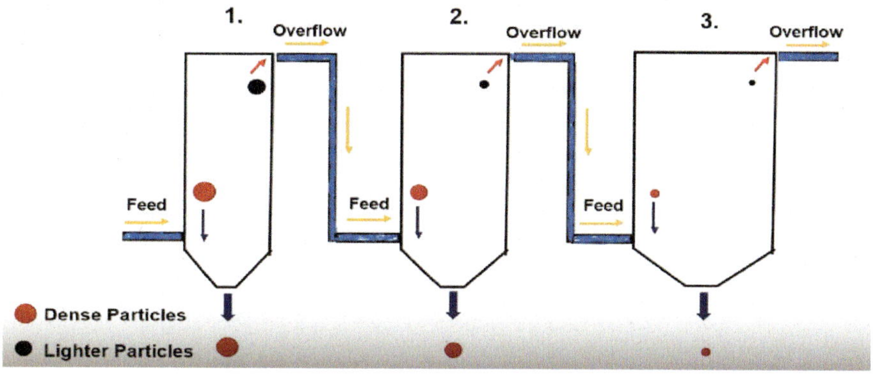

Elutriation method of separating the different size soil particles

(Source: https://www.youtube.com/watch?app=desktop&v=Djo4hmkuieI)

d) **Sedimentation method**: This method is based upon the settling velocity of the particles suspended in water. It is used for finer particles which cannot be separated by sieving.

Here soil suspension is made in distilled water without dispersion and followed in the same way as particle size distribution.

Expression of result of aggregation

Mean weight diameter (MWD): "It gives an estimate of the weighted percentage of average sizes of all the aggregates".

$$MWD = \Sigma X_i W_i$$

Where Xi = mean diameter of any particular size range of aggregates

Wi = weight of aggregates in that size range

The difference between MWD of dry sieving and that of wet sieving is called the change in mean weight diameter (CMWD). The lower value of CMWD indicates the better structure.

Chapter 10

Soil Consistency (Rheological Property)

VISCOSITY

Viscosity is a very important property of liquid which describe their flow under resistance force. "Viscosity is defined as the resistance of the liquid to its free flow due to internal friction between their layers as they pass over each other with different velocities".

It is the property which opposes the relative motion of adjacent proportions of liquid and can be regarded as the type of internal friction. Viscosity of a liquid is a measure of this internal friction of the liquid and it determines the rate of flow of liquid.

Jean Luis Marie Poiseuille (1844), the French physicist found that if a liquid with a coefficient of viscosity (η) flows with a uniform velocity @ V cm^3 in t second through a narrow tube of radius R cm and length L cm through a driving pressure P and pressure drop ($P = P_1 - P_2$), then

$$\eta = \frac{\pi \Delta P R^4 t}{8VL}$$

$$Q = \frac{V}{T} = \frac{\pi \Delta P R^4}{8\eta L}$$

where Q is the discharge in cm^3/hour

This equation is known as *poiseuille's law* which state that the volume of water flowing in a unit time (Q) is proportional to the pressure drop per unit distance (P/L) and fourth power of the radius of the tube.

The units of viscosity are dyne second/sq cm which is also known as poise. One poise is equal to 100 centipoise. The viscosity of water at 20°C is 1 centipoise.

IMPORTANCE OF VISCOSITY

Viscosity has a crucial role in finding the

- Rate of fall of a soil particle through the viscous fluid
- Rate of upraise of water bubble
- Rate of flow of water through a small narrow tube

So, fluid viscosity is used for estimating the different soil separate fractions during analysis of soil for texture. Temperature decrease the viscosity of the soil solution due decrease in internal cohesion or intermolecular attraction. Thus, low coefficient of viscosity favours the liquid to flow readily with little or no resistance force while liquids with high coefficient of viscosity are relatively less mobile as they are acted upon resistance force that restrict their free flow.

FACTORS AFFECTING VISCOSITY OF SOIL SOLUTION

1. **Temperature:** Increasing temperature decreases the viscosity of the soil solution because on rising temperature the average thermal energy of its molecules increased. This happens as the internal friction due to intermolecular force of attraction decreased, thus viscosity decreased.
2. **Electrolyte concentration:** The inclusion of small amount of electrolyte to a soil colloidal system lead to a sharp drop of their viscosity.
3. **Surface charge:** The surface charge on the particle affects the viscosity of the soil solution. Thus the soil solution with minerals possessing high surface charge has high viscosity. Eg. Montmorillonite clays possess more surface charge than that of kaolinite because of extremely small size of their particles, thus the intermolecular attraction of molecules increased leading to increase the viscosity of the solution.
4. **Type of clay minerals:** Variation in viscosity is due to difference in hydration of the molecules of the solution. The expanding type clays such as montmorillonite are more hydrated than non-expanding type clay minerals like kaolinite.
5. **Exchangeable cations**: The presence of large hydrated radius ions like that of of Na will increase the viscosity of the soil solution while that of ca and mg will decrease the viscosity as the former leads to dispersion of the soil while the later flocculate the soil particles.

Soil consistency: Rusell and Rusell (1950) define "soil consistency as the strength with which soil materials are held together or the resistance of soils to deformation and rupture at different moisture content. It is the ability of the soil to change the shape or to be moulded".

Two main forces responsible for soil consistency are

a) Cohesion (molecular attraction)

b) Adhesion (surface tension)

Cohesion (molecular attraction): It is the force that allows the substance with similar characteristics to attract each other for eg. attraction of a water molecules with other water molecule or of one soil particle for another soil particle. The cohesion force (molecular attraction) is due to zeta potential or van der waals forces of attraction, surface charge of the particles and the broken edges of the clay minerals/ particles. The cohesion force is large when the soil particles lie close together having relatively large surface area in common, as in case of oriented colloidal clay. The cohesion force is greatest in dry soil and it decreases as water enters between the soil particles and keeping the particles apart. The loose dry soils have no cohesion. As the water of the loose dry soil increases, cohesion increases rapidly to a maximum value and then decreases due to thickening of water films between the particles and keep them apart. In case of puddled soils, cohesion increases with decrease in water content as it lead to lesson the thickness of water films around the clay particles.

Adhesion (Surface tension): It is the force that allows the substances of dissimilar characteristics to attract to each other probably due to surface tension forces. The adhesion force is highest when the meniscus is curved inwards the most i.e. moisture content in soils is in great tension or water is tightly bound to the soil particle. At this moisture content, the cohesion force decreased due to decrease in the contact area of the soil particle. With the further increase in the water content of the soil, the surface tension force also decreased, thus consistency decreased. At particular moisture in wet range, the decrease in adhesion force cannot compensate for the increase in cohesion force, resulting decrease in consistency. Under saturated condition, the cohesion and adhesion forces become zero, thus consistency decreases. A plot of consistence against moisture status of a soil would show two maxima and two minima. Consistency is large at the extreme low level of moisture due to cohesion and in the middle of the wet range due to adhesion.

These relationships hold good in soils with sufficient clay, as silt and sand contents contribute little to molecular attraction. Once the consistency of a clay soil is destroyed and the soil particles are separated, it can only be re-established after wetting the soil for re-orienting its particles again. Other factors that may affect consistency are:

1. ***Type of clay***: The montmorillonite clay possess more surface area than that of kaolinite clay minerals, thus the former shows more consistency than the latter.
2. ***Texture***: Clays due to its extremely small size shows more specific surface area than sand and silt, thus consistency of the soils with higher clay content is more than the soils with higher proportion of sand and silt.
3. ***Organic matter***: Organic matter due to its friable properties has more cohesive force than the coarse fractions of the soil i.e. silt and sand but

less than that of finer fraction of the soil i.e. clay. Thus, organic soils have low consistency than that of clay soils.

4. ***Structure*:** A soil with high bulk density as that of the puddled rice soil has more consistency than the soils with well-aggregated structure as puddled soils possess higher contact areas between the particles. The disperse soil due to its more area of contact has higher consistency than that of aggregated or flocculated soil.
5. ***Presence of sesquioxides and Oxides and hydroxides of Ca and Mg*:** The presence of sesquioxide and Ca and Mg oxides and their hydroxides will increase the consistency due to more power to form floccules.

Forms of soil consistency:

1. ***Hard or Harsh consistency*** (pF > 4.5)**:** This consistency is noticed in dry soil. At low moisture content, the soil remains very hard and coherent due to cementation effect between the dried particles. If tillage implements are carried out at this consistency, it will form clods. Heavy draft is required at this consistency.
2. ***Soft and friable consistency*** (pF = 4.5 to 2.8)**:** This type of consistency is noticed in moist soil. As moisture content of the soil increases, water molecules are adsorbed on the surface of the soil particles and decrease the coherence. Soil particles stick together only slightly. The soil mass becomes friable. Friability characterizes the ease of breakdown of soils. The range of soil moisture contents in friable condition is optimum for tillage operations. Draft required is less at this consistency.
3. ***Plastic*** (pF = 2.8 to 0.5)**:** This form of consistency is found in wet soil. Soils can be changed in to desired shape and will regain its original shape as the pressure is released or lessoned, without losing its coherence. At this consistency soils can be easily puddled and plasticity is the main characteristics of clay. Heavy draft is required at this consistency. Implements tend to sink and slip and soils are easily puddled.
4. ***Sticky*** (pF = 0.5 to 0)**:** This consistency occurs in a highly wet soil. At this consistency, soil adheres to other objects.
5. ***Viscous consistency*** (pF = 0)**:** This form of consistency is for saturated soils. At this consistency, soil flows either due to pressure or merely by gravity. At this consistency, soil runs together and tillage is impossible.

Factors affecting soil consistency

a) **Soil water content:** The soil consistency is due to cohesion and adhesion, so it is large in dry and wet range. It is small in moist and saturated conditions.

b) **Soil texture:** As sand and silt have no molecular attraction, consistency of coarse textured soil is negligible. However, soils possessing considerable amount of clays have higher consistency.

c) **Type of clays:** Amount and type of clay colloids greatly influence the soil consistency. Expanding type of clay minerals because of small size particles and high surface charge exhibits more consistency than non-expanding type clay minerals.

d) **Organic matter**: It has higher consistency than silt & sand but less than clays because of its higher cohesion force than that of cohesion force existing between sand & silt and less than that of cohesive force of clay.

e) **Structure**: Well aggregated soils has less consistency than a puddled soils because well aggregated soils has less area of contact between the individual soil particles than the puddled soils. When puddled soils become dry, its consistency sharply increased because of increase in surface contacts.

Soil plasticity: Soils with clay percentage of < 15 will not show plasticity phenomenon in any wet range. This amount, however, depends on the soil's type & amount of clay minerals and its organic matter content.

"Plasticity is defined as the property which enables a clay/soil to take up water to form a mass that can be deformed into any desirable shape and to maintain the shape after the deformation pressure is removed".

Plasticity may be due to the platy type of structured particles and the strong binding properties of water. The adsorbed water will form thin films along the soil particles. As the pressure is applied in the moist soil, the particles slip on each other and the water is held on to the soil as a thin film even when the pressure is released.

Indices of plasticity and their significance: (Atterberg 1911-12) The three indices along with their significance is explained as below:

***Lower plastic limit or the plastic limit*:** This limit represents that moisture content at which soils can be easily changed in to desired shape without being ruptured or form cracks. This moisture content is the upper limit where tillage implements can be carried out for many crops, except transplanted/puddled rice. There may be smearing or puddling of the soils if the tillage is done at moisture content above this limit and below which soils act as soft and friable.

This limit is determined by making the soil solution in a round bottom container by adding water up to the point when it lose its crumbly feel and tend to behave like a plastic. This moist soil is taken in the hand and is allowed to roll between the fingers and palm until it form the wire. This process is repeated until the moisture content is reached when the mass of the soil just barely rolls out into a wire that breaks into pieces.

***Upper plastic limit or liquid limit*:** The upper plastic limit or liquid limit represents the moisture per cent at which a soil stops to behave as plastic as because the soil becomes semi-fluid and flows like a liquid when little pressure is applied. This limit is being used by the engineers for classification of soils.

Casagrande apparatus

(Source: https://www.vssut.ac.in/lecture_notes/ lecture1625036760.pdf)

The upper plastic limit is estimated by taking the small amount of soil in a petri dish and making a paste of it, and press it tightly, cutting a V-shaped groove in the plastic mass and jerk the round dish to make the two section to flow together. If there is no flow, add some extra water in to it and the process is repeated. If the flow is very high then mix the dry soil with the plastic mass. This process is continued till the correct and free flow is reached. This way the moisture of the soil is determined. The apparatus used for determining the liquid limit is called as *Casagrande apparatus*.

***Plasticity index*:** The range of the moisture content over which soils behave as plastic is called plastic limit. High plasticity indexed soils are not easy to till. Soils have high amount of montmorillonite and vermiculite clay minerals whioch expands on wetting have higher liquid limits and plasticity indices.

While the soils with high amount of Kaolinite clays shows lower liquid limit values.

Soil plasticity is determined by an apparatus called ***liquid limit device***.

The value of plasticity index estimates the capacity of the soil to compress. The higher is the plasticity index, the higher is the compressibility of the soil.

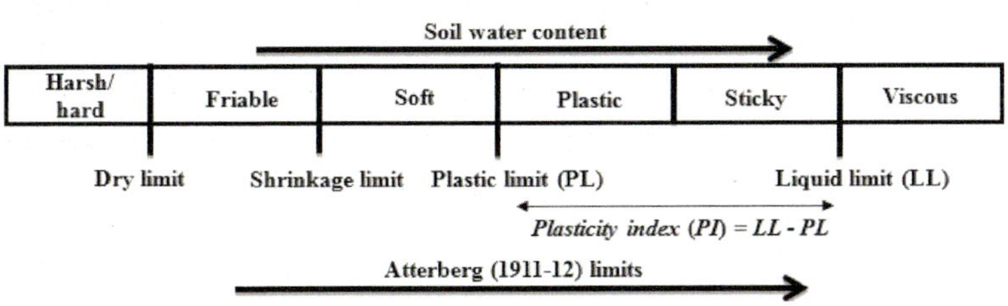

Atterberg's limits are being used by soil engineers for checking the suitability of soil for their construction purposes. Higher plasticity index representing > 25, results in poor road beds or foundations.

Coefficient of linear extensibility (COLE) can be used to quantify the soil's expansion nature. For eg. If a sample of soil is moistened to its plastic limit and molded into the shape of a bar with the length say L1 and it shrinks to length L2 when the bar of moist soil is air dried, then COLE can be calculated by finding out the per cent decrease in the length of the soil bar upon drying in air as follows:

$$\text{COLE} = \frac{L1 - L2}{L1} 100$$

Laws of plastic flow

Plasticity is the effect of stress and deformation without being rupture or cracking when it is subjected to external force at various moisture contents. Therefore, some stress is needed to produce plastic flow.

Plastic flow in a soil can be described by the 'Bingham equation':

$$V = K \mu (F - f)$$

where, V - volume of flow,

μ - coefficient of mobility,

F - force applied and

f - force necessary to overcome the cohesive force in the system

The yield value f is obtained by extrapolating the straight line segment to the intersect point on abscissa. The magnitude of yield value depends on the extent of cohesive forces of the water films between particles.

Sands when wet may be moulded slightly but the moulded form breaks into small pieces when dried. Therefore, sands are not plastic. The soils less than 14-16 % clay will not possess plasticity in any moisture range.

Law of viscous flow

Bingham equation may also be used to describe the viscous flow if $f = 0$ i.e. for viscous follow the equation is

$$V = K\mu F$$

Therefore, for the viscous flow, the volume of flow is directly proportional to the force applied and coefficient of mobility of the liquid (Fig)

Chapter 11

Soil Crusting

Soil crust: A soil crust may be surface or sub-surface which can be defined as the compacted and hard layer of the soil than the immediately underlying soil on drying. The crust may be several mm to several cm thick depending upon the clay minerals present (type and amount), and also its silt content. The soils having higher proportion of clays have thicker crust. So, "*soil crust is defined as the formation of compact thin layer having high bulk density and high penetration resistance at the surface of soil when dries after dispersion of fine particles and clogging of soil pores due to beating action rain drop and slaking action of irrigation water*".

Structural crust is the major structural feature of soils of arid and semi-arid regions where crust formation is mainly due to destruction of aggregates caused by impact of rain drops or irrigation. Crust may be formed on all types of soils except sand. Silty clay loam alluvial soils are most susceptible to the formation of crust. It is characterized by high bulk density, very low non-capillary pores, very low hydraulic conductivity and high penetration resistance.

MECHANISM FOR THE FORMATION OF CRUST

The soil aggregates with low water stability, break/disintegrate and get dispersed due to the impact of rain drop falling on it or due to high intensity of sprinkler irrigation. The soil particles so dispersed comes in to solution form. Coarse particles due to its bigger size and more weight start to settle first but fine particles remained suspended in water for longer period of time. After certain period of time as the water drains off or gets evaporates, clay starts deposits on the top of coarser particles which on drying forms the surface crust which become difficult for the plant roots to break.

The soil particles so dispersed enters into the pores spaces and clog it due to which layer becomes dense with high bulk density. This in turn will reduce the infiltration of water, its percolation, restrict drainage, and restrict the movement of nutrients, heat and air and also exchanges of gases from soil to atmosphere and vice-versa. The layer turns hard as it dries and further gets harder with more strength with further decrease in its moisture content. Besides moisture content, hardness also depends upon the oxides of Fe and Al, $CaCO_3$, and silt content. Higher are these components in the soil higher is the strength/hardness of the crust.

The important soil properties that encourages the formation of soil crust are low organic matter content, high silt content and high exchangeable sodium content of the soil.

Types of soil crust

Soil crust can be broadly categorized under two heads:

A. Structural crust

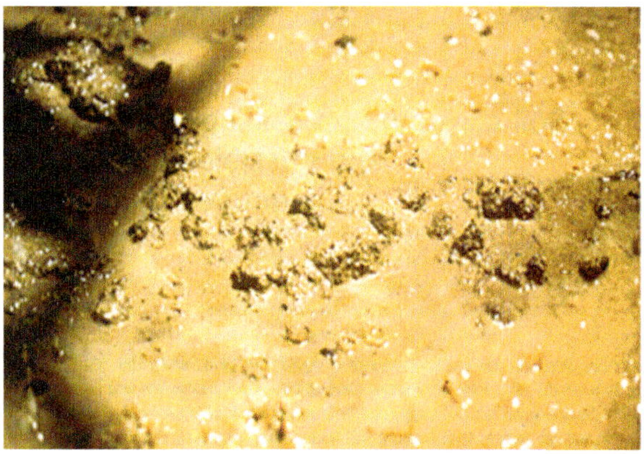

Slaking crust
Source: https://www.fao.org/3/T1696E/t1696e10.htm

a) **Slaking crust**: When the high intensity rain on falling to the soil disintegrates the weakly aggregated soils or the sodic soils, the crust is formed which is called as slaking crust. These crusts made up of thin layers of fine particles with broken clods. Slaking crusts are formed when soils contain high i.e. > 15% clay content. to entrap and compress air during wetting so that aggregates break down. Since slaking crusts are formed due to wetting of the soil mass, even without the effect of high intensity rainfall. The textural differences between sand and clay particles are indistinct or not visible. These crust are more porous than the depositional crust.

b) **Infiltrating crust:** This crust developed on medium textured loamy soils due to effect of rain drop eroding the aggregates lying on the surface causing the silt particles to seal the voids/pores spaces. They are mainly characterized by clear textural separation.

c) **Coalescing crust:** These types of crusts are developed in very wet soils when the soil material is in viscous state by the drop energy driving force. These crust are thicker in depth (up to 20 mm), packed densely than the underlying layer,' form as a transition layer from the underlying layer due to which they are also termed as transitional micro horizons. This crust has less porosity at the top layer but the pororsity gradually increases with depth.

d) **Sieving crust:** Sieving crusts are formed when the loose sandy particles overlays the thin layer of clay/fine material. In this type of crust, the three different layers exists which are well sorted. The inner most or the lower layer consists of very fine particles, overlying this layer is the fine dense packed grains of soils with vesicular pores while the top most consists of particles of loose sand which is coarser. These three layers lying one over another reduces the porosity of the soil.

This lower layer causes the low infiltrability (0-15 mm h^{-1}) of the crust. The sieving crust leads to differentiation in texture of the soil. Also, the clay particles move to the lower layer through upper sand layer through percolating water or by wind. Clay particles can accumulate due to entrapped air within the underlying layers during infiltration. This type of crusting mostly observed in sandy and sandy-loam soils.

B. Erosional or depositional crust

a) Erosional Crust:

Erosional crusts are formed when the upper loose sandy textured layer of sieving crusts is eroded by the runoff water disclosing the underlying layer of fine clay particles at the surface. These erosional crust are thin ranging from 100 pm to 1 mm thick, plasmic layer which is very dense, smooth and coherent mass of very fine particles of clays. Porosity is sharply reduced leading to reduced infiltrability of water. The erosional crust are formed from slaking as well from sieving crust. The slaking crust in erosional crust are smoothed and further enriched with more fine particles whereas sieving crust by removing the upper coarsed textured layer soil.

b) Depositional Crusts

These type of crusts are formed when structural crust seal the surface of the soil and the particles from eroded clods and ridges gets detached and concentrated as a micro deposition between the interclods and in the furrows during runoff. Thus, these crust form a sedimentary layer of 5 to 10 mm thick on the structural crust formed previously. The material of the layers in deposition crusts are the reverse of that occurred in the sieving crust with finer clay particles at the top of the layers and coarser sand particles at the bottom. Thus, again the texture differentiation is distinct or visible.

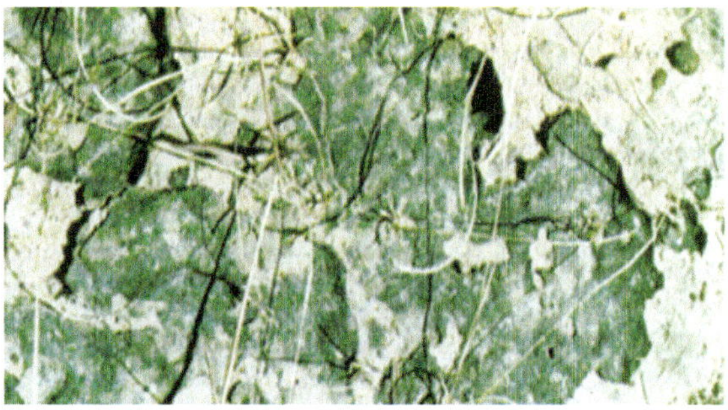

Depositional crust

Source: https://www.fao.org/3/T1696E/t1696e10.htm

This distinct differentiation of texture resulted in the development of microbials which are less distinct in the lower part of the sedimentary crust layer.

C. Cryptogamic Crusts:

These crust are formed by microbes such as algae, fungi, lichens, mosses, bacteria over the structural and erosional/depositional crust. They are also known as microphytic crusts as they consist of a microbiological layer over one of the above defined crusts. These crusts are not always considered as soil crust due to multiplication of microorganisms over the structural or erosional or depositional crust.

Adverse effects of crusting: The soil properties likely to be affected due to formation of crusts are bulk density, infiltration rate of soil, thermal conductivity and diffusivity and surface runoff. This will adversely affect the germination of roots from seed due to high emergence force required to break the crust, plumule formation and emergence of seedling. This inturn will depend on crust thickness and its strength and the package of practices adopted for its management to get rid of adverse effects of crusting.

Soil crusting is more prone in arid and semi-arid regions which cause high surface runoff, thus leaving less water for plant growth.

Measurement of crust strength: The strength of crust can be measured by either employing modulus of rupture test or through use penetrometer which measures the resistance force implied to penetrate the soil.

Suppose if the maximum crust strength of a sandy loam soil can be about 3.5 kg cm^{-2} with a crust of 3.0 mm thick, then the emergence force depending on specific crop is as follows:

Crop	*Emergence force (kg cm^{-2})*
Bajra	2.1
Cotton	3.1
Moongbean	3.3
Clusterbean	3.9
Maize	16.5

Thus, soil crust becomes a limiting factor for the emergence of Bajra, cotton and mung-bean especially in the dry regions prevailing sandy loam texture which has an emergence force of less than the crust strength of 3.5 kg cm^{-2}. This resulted in poor germination with less plant population. Once the crust is formed after sowing of the crop, then the resowing of the crop can be done. The adverse effect of crust can also be minimized by growing the crop on ridges.

Measures to be adopted for the control of crust formation: Crust formation can be reduced by:

1. Mulching either vegetative or plastic can be used to minimize the impact of rain drop directly to the soil.
2. Certain management practices such as use of soil organic matter and use of certain natural or synthetic soil conditioners like hydrolyzed polyacrylo-nitrile (HPAN) or vinyl acetate maleic acid copolymer (VAMA); nonionic soil conditioners like polyvinyl alcohol (PVA) etc. can reduce clay dispersion and crust formation.
3. Application of light tillage will break the crust before it becomes hard.

Control measures for mitigating the adverse effect of crust on emergence

1. Covering the seed line area with FYM or straw just after sowing of crop protecting the soil from direct impact of rain.
2. Sowing on ridges helps in increasing the emergence of the seeds.
3. After the formation of soil crust, scraping the surface soil with a tined hoe can increase the crop emergence.

SOIL COMPRESSION AND SOIL COMPACTION

*"**Soil compression** is the term used to describe the reduction in volume of a soil when an external source of pressure is applied"* whereas *"soil compaction refers to the increase in density of a soil as a result of applied pressure"*. When a pressure is applied on to the soil, it may result the decrease in its void ratio which may occurs due to reorientation and reduction in size of the colloidal particles. After removing the pressure or when the stress arises due to certain factors releases, the soil expands due to presence of swelling type of clays. The factor that may affect the compression are

1. Type and nature of clay mineral and
2. Exchangeable amount of cations

Compression increases up to a certain increase in moisture content and thereafter it decreased as the moisture content increased further. The soils can be compressed to a maximum within the plastic range of soil moisture content.

"**Soil compaction** includes compression with further increase in bulk density of a soil". The compaction of soils depends upon the following factors

1. Texture of soil
2. Extent of energy/pressure applied
3. Content of water present
4. Extent of manipulation of the soil.

Soil compaction is also associated with the rearrangement of the soil solid particles so that soil is compressed within pore spaces. Dry soils have high internal friction and incompressible in nature thus cannot be compacted to much intensity. Cohesion and internal friction force between the particles decreased with increase in moisture content, thus applying pressure/load will facilitate compaction. With further increase in moisture content of the soil, the cohesion as well as adhesion force decreased resulting decrease in compaction due to incompressible nature of soil and water. This critical limit is called Procter moisture content.

The compaction soils in most soils will occur at below the plough layer, thus is termed as plough sole. Compacted surface or subsurface layer resulted in poor root growth due to poor root penetration and reduced nutrient and water uptake by the crops. The soil compaction can adversely affect the growth of most of the crops in fine textured clayey soil but it could be useful in coarse textured sandy soils as it may help in retaining water and reduce the percolation losses of water.

Soil compaction	Soil consolidation
1. Instant process	1. Slow process
2. Artificial process	2. Natural processq

Soil compaction	Soil consolidation
3. Compaction is due to expulsion of soil air under short duration	3. Consolidation is the expulsion of pore water under steady state, static and long term load
4. Water content will remain the same and density will increase	4. Water content will decrease with increase in density
5. Compaction is of agricultural significance	5. Consolidation is of interest for civil engineer for foundation and construction of highways

Factors affecting soil compaction

a) Soil-water content:

- Dry soils cannot be compacted due to high internal friction between the particles and also the incompressible nature of soil particles.
- As soil water increases, thickness of water film around the particle will increase thereby reducing cohesion and thus the soils can be compacted to higher degree.
- The increased compaction of a soil with increase in water content under constant pressure is due to
  - Lubricating action of water
  - Reduction in cohesion between particles
  - Orientation of soil particles close to each other due to expulsion of soil air
- With further increase in water content, compactness decreases again

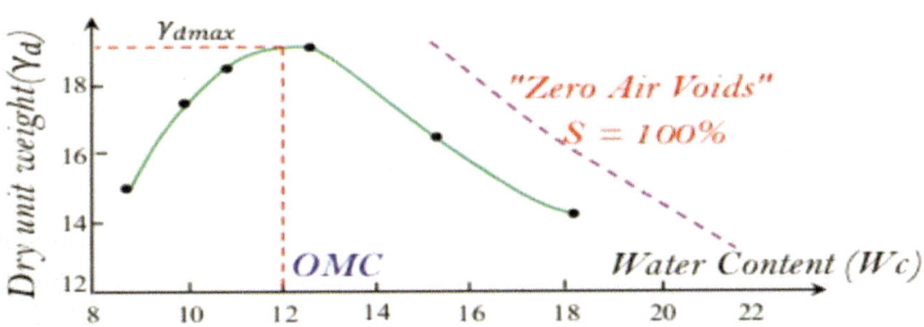

b) **Soil texture:** Cohesive soils such as heavy clays & silts provide higher resistance to compaction as they attain lower maximum dry density. Cohesionless soils such as sandy soils and coarse grained or gravelly soils are susceptible for easy compaction.

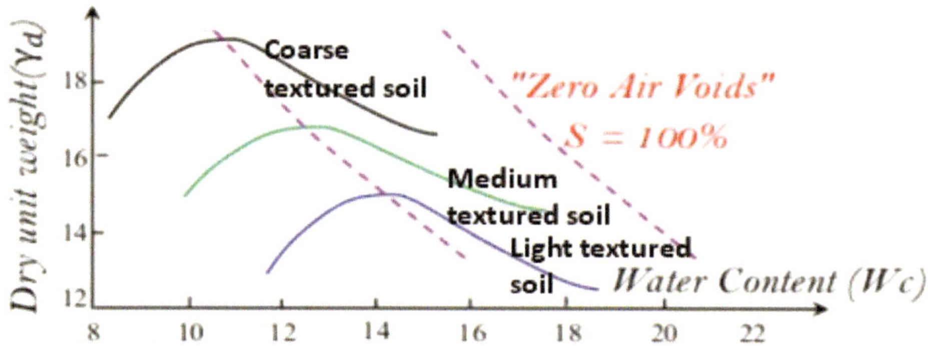

(Source: https://environment.uwe.ac.uk/geocal/soilmech/compaction/compaction.htm)

c) **Amount of manipulation:** Soil compaction causes reduction in total pore space especially air filled pores without change in mass of the soil solid. Continuous compaction will remove total air filled pores causing removal of all air. If compaction is continued further, there will be complete reduction of pore spaces resulting in expulsion of air and water

SOIL COLOUR

Soil colours have little influence on the behavior of soils but provide information regarding processes occurring in the soil according to its behaviour. It is very important in soil classification, interpretation and description of soil profiles, the processes involved in the soil formation, presence of excessive salts, erosion, etc. Colour of the soil indicates the moisture content under which it was developed.

Determination of soil colour: ***Munsell colour chart*** represents the standard system for describing the soil based on its colour. The soil colour is compared to the colour chips of Munsell colour chart which was described by the parameter –

i) Hue,

ii) Value and

iii) Chroma.

"The *hue* is defined as the dominant spectral colour and refers usually to redness or yellowness or greenness or bluish colouration of the soil. The *value* refers to the relative lightness or darkness of colour, a value of 0 being black. The *chroma* is the relative intensity or brightness of the dominant colour, a chroma of 0 being a neutral grey". Soil colour indicates the oxidation and reduction processes, presence of iron containing minerals, the water content of the soil or drainage condition of a soiletc.

The high croma soils with bright colours such as red colour show that the soils have good drainage and are well aerated. While soils with waterlogging condition for considerable period of time or poor drainage reduces oxides of iron oxide and their coatings and changes to low chroma i.e. grey, bluish or grey-green colouration.

Presence of low chroma colour soils i.e. greyish colour either as a whole or presence in motlled form with various tinges of gray, brown and yellow is the criteria to identify imperfectly and poorly drained soils or water logged conditions for at least maximum part of growing period of the crop.

Soils may differ widely as per the colour viz. red, brown, yellows, black, white and even green. Soil colours may vary from place to place in the landscape. Soils adjacent to each other may even differ in surface colour e.g., black and red soils existing side by side in Andhra Pradesh and Maharastra. The colour in a soil profile may change with depth through the different horizons. Sometimes the horizon resembles in colour due to same hue but differ due to different values of chroma and value.

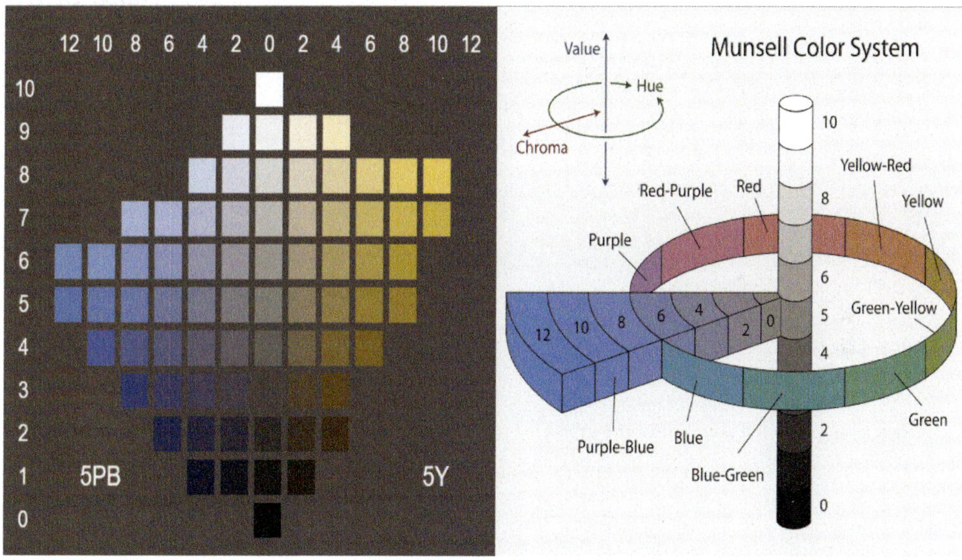

Munsell Colour Chart

(Source: https://en.wikipedia.org/wiki/Munsell_color_system)

Factors influencing soil colour

1. **Mineralogy and chemical constituents of soil particles**: Red, yellow or brown colours are indicative of the extent of oxidation process, hydration of iron minerals and diffusion of iron oxides in the soils which form coatings on the soil particles. Thus, yellow, red and brown colours are mostly due to the presence of iron bearing minerals viz. goethite, hematite and magnetite, respectively.

2. **Organic matter content**: Presence of organic matter tends to make the soil dark brown to black. As a consequence, soil horizons high in organic matter are darker in colour than those of low in organic matter. Also, the conversion of organic matter to Humus makes the soil colour more darkers than before.
3. **Soil moisture**: Moist soils are generally darker than dry soils. Water content of a soil also indirectly affects the soil colour through its influence on soil oxygen.
4. **Soil texture and structure:** Soils of dry regions are lighter in colour due to low content of organic matter and less water as compared to soils of wet regions which has generally high organic matter content. Forest soils are darker than the grass land soils due to high amount of humus and organic matter content and also high extent of microorganism activity in the soil. Sodic soils on drying shows darker surface crust layer due to its high organic matter. A strong soil structure may provide sufficient aeration that may lead to differing colours of soil ranging from yellow to red and brown depending upon its mineralogical composition. However, a poorly structured soil has restricted aeration and poor drainage condition due to which the colour of the soil may form greyish green.
5. **Oxidation states of iron and manganese oxides present in the soil:** The oxygen determines the oxidation states of several elements, especially, of iron in its oxide forms. Minerals like manganese oxides and glauconite impart black and green colour, respectively to soils. Carbonates, such as calcite, dolomite accumulating in soils of semi-arid regions impart a white colouration to soil due to deposition of salts at the surface. Under prolonged anaerobic conditions, the coatings of the reduced iron oxide on soil particles change the soil colour to grey or bluish.

Chapter 12

Soil Water: Properties, Classification and its Relationships

Water supply through soils is vital for both plants and soil organisms—they need water to survive. Soil water contains nutrients that move into the plant roots when plants take in water. Water enters the soil through large pores (macropores) and is stored in many small pores (micropores). Porous soils have a balance between macro and micro pores. Also, the water supplied to plants through soils is continuously lost by evaporation and transpiration. Water is the solvent through which nutrients in dissolved form is supplied to plants for its normal growth and nutrition. Soil water helps to control soil aeration and heat. Water is the most important component of soil which is most dynamic and influences all processes occurring in soil and plant system

Soil water is the liquid phase of soil which is composed of dissolved salts and gases, therefore, generally called *as soil solution or soil moisture.*

STRUCTURE OF WATER

Water due to its structure directly and indirectly takes part in many soil and plant reactions. The chemical formula of water is H_2O i.e. each molecule of water contains two very small atoms of hydrogen and one atom of oxygen. Hydrogen is bonded with oxygen with a strong covalent bond. The sharing of electrons in water molecules is arranged in tetrahedral structure with two hydrogen atoms share with one oxygen atom with a strong covalent bond.

The arrangement of electrons in the water molecules creates an electrical asymmetry. However, the two hydrogen atoms are both on the same side of the oxygen atom so that positive charged nuclei of the hydrogen atoms are left exposed, thus leaving that end of water molecule with a weak positive charge. While on the other side of water molecule, the excess electron of the oxygen atom gives that end of a molecule a weak negative charge. For this reason water is called a 'dipolar molecule'.

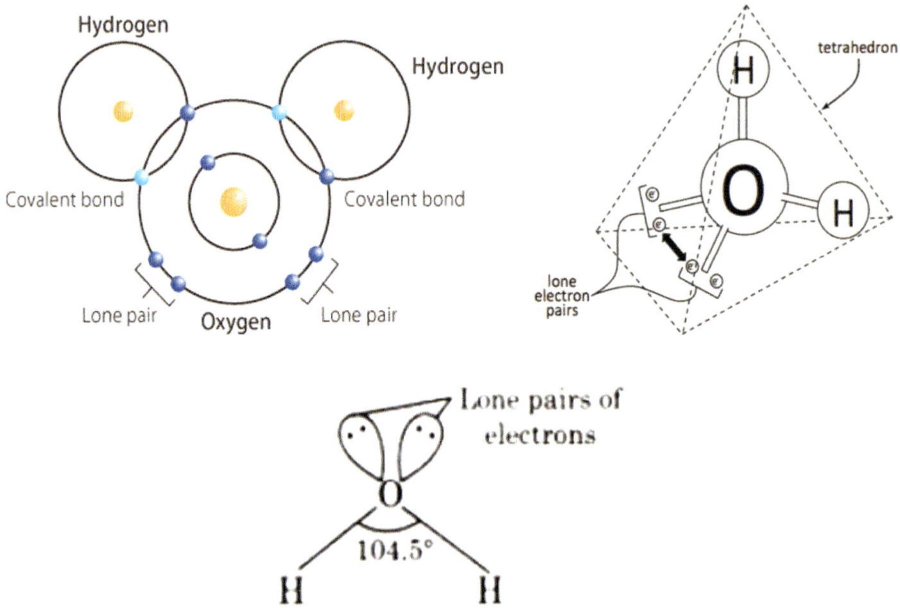

Bent structure of water molecule

(Source: https://www.expii.com/t/water-molecular-structure-bonding-10383

In solution, the weak positive charge side of one water molecule will be attracted to weak negative charge side of the another water molecule and two molecules of water are held together by a weak hydrogen-bond (H-bond). In the liquid state, a molecule of water is surrounded by five or more water molecules, while in the solid state; each of the water is linked to four other molecules in a hexagonal type arrangement. This makes water lighter in solid state than in a liquid state.

The weak H-bond are constantly being broken and reformed. This gives water some structure but allows the molecules to slide over each other easily making it a liquid. In addition to shared pairs of electrons, the oxygen atom of the water molecule consists of two lone pairs. These two lone pairs of electrons do not takes part directly in bond formation as do electrons shared between hydrogen and oxygen atoms of the water molecules. The electrical structure of the water molecule resembles a tetrahedron with oxygen near centre, two of its corner positive charge and remaining two corners negative charge due to two lone pair electrons.

Thus, this polarity (unequal distribution of charge) of the water molecules makes them mutually attractive, makes the water to adsorb on the soil surface, hydrate ions, capable of dissolving most other compounds and acting as a good solvent. Water is angular in shape; the two hydrogen are separated from each other at an angle of 104.5 °. The lone pairs are closer to the oxygen atom than the electrons bonded to hydrogen, so they require more space. The increased repulsion of the lone pair forces O-H bonds closer to each other. Due to this, water has a best structure.

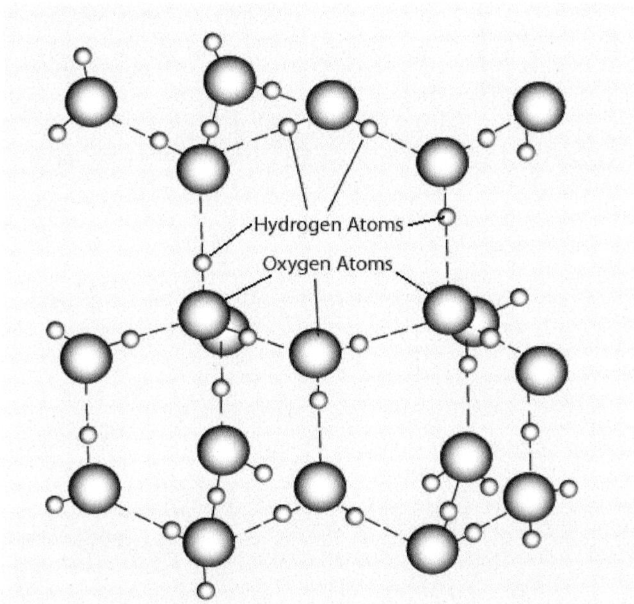

Hexogonal structure of solid water
(Source: https://www.bartleby.com/)

Due to H-bonding, liquid water has higher values of melting point, boiling point, dielectric constant, specific heat and viscosity as compared to other compounds of similar molecular weight.

PROPERTIES OF WATER

A. Density

The open structure of the water molecules is the cause of its relatively low density. Water exhibits maximum density at 4°C (1.0 Mg m^{-3}). At 0°C temperature the density of the water decreases to 0.992 Mg m^{-3} which may be due to expansion of water due to hydrogen bonding and water forming hexagonal structure with complete bonding having open spaces between the molecules. However, the density of water molecules decreases to 0.997 Mg m^{-3} with increase in temperature to 25°C. This may be due to expansion of water due to increase in kinetic motion of the molecules for which each molecule takes greater space. The decrease in density of water is very small with increase or decrease in temperature due to which density of the liquid water is generally taken as 1.0 Mg m^{-3}.

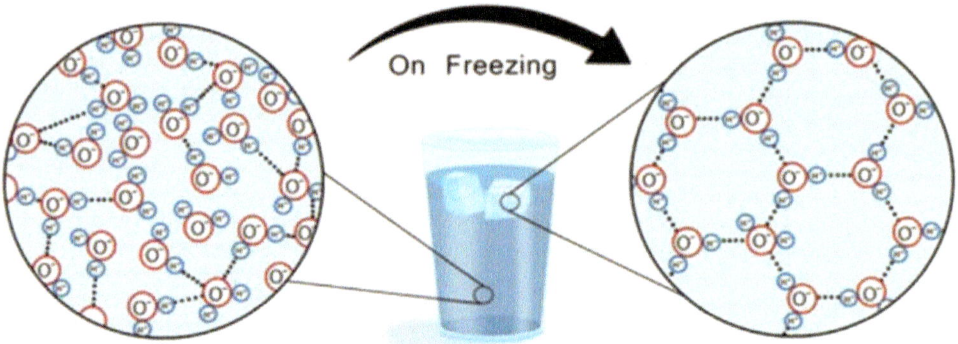

Liquid water **Solid water**

(Source: https://www.quora.com/)

The difference in density of liquid water and ice is large. As hydrogen bonding is complete in ice, its density is considerably lower (0.917 Mg m^{-3}) than liquid water. An approximate increase of 9.0% volume was observed when water has converted to ice. In liquid state, water molecules are closely held together by weak hydrogen bonds which on reaching the solid state become stable, arranging water molecules far apart from each other. Thus ice 's density becomes lesser than water which makes it to float on water.

B. Surface tension

Surface tension is due to an imbalance between intermolecular attractions (cohesive forces) at the surface of the water. Water molecules hold each other due to the intermolecular cohesive forces. These forces, acting on a water molecule are effectively equal in all the directions in the bulk solution. Thus, the net force on the molecules in the bulk solution is zero. However, a resultant inward force acts on the molecules at the air-water interface (surface) due to the absence of the water molecule above. The presence of fewer water molecules at the surface also results in a stronger bond between the molecules. This inward net force causes molecules at the surface to contract and to resist being stretched or broken. Thermodynamically, the molecules at the surface must gain excess energy over those in the bulk water to retain at the surface. This surface free energy is the surface tension at the air-water interface and the surface behaves like a tension membrane.

The units of surface tension are energy/area (ergs/cm^2) or force per distance (dynes/cm). The surface tension of water is very high (72.7 dynes/cm at 20 °C) as compared to other liquids like ethyl ether (17 dynes/cm), ethyl alcohol (22.0 dynes/cm) etc. This higher surface tension of water molecules is due to higher attraction of water molecules for each other.

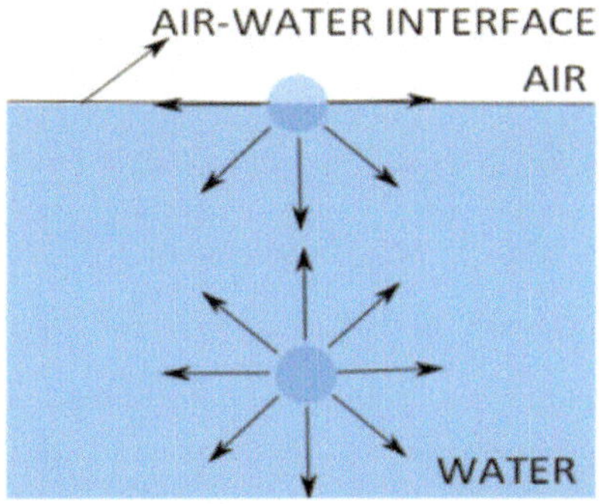

Surface tension at air-water interface

(Source: https://archive.nptel.ac.in/courses/105/103/105103139/)

Surface tension of the water generally decreases with increase in temperature due to thermal expansion which tends to decrease the density of water and therefore, reduces the cohesive force at the surface as well as inside the liquid phase. Addition of electrolytes may increase or decrease the surface tension force of the water. Inorganic solutes like sodium chloride increase the surface tension of the water while organic solutes like detergents decrease the surface tension of water due to more cohesive attraction between water molecules than their attraction to solute molecules.

C. Contact angle of water on solid surface

The contact angle is the angle, conventionally measured through the liquid, where a liquid–vapor interface meets a solid surface. It quantifies the wettability of a solid surface by a liquid via the Young equation. A given system of solid, liquid, and vapor at a given temperature and pressure has a unique equilibrium contact angle.

If the solid–gas interfacial energy is denoted by γSG, the solid–liquid interfacial energy by γSL, and the liquid–vapor interfacial energy (i.e. the surface tension) by γLG, then the equilibrium contact angle θ is determined from these quantities by the Young equation:

$$\gamma_{SG} - \gamma_{SL} - \gamma_{LG} \cos\theta = 0$$

$$\cos\theta = \frac{\gamma SG - \gamma SL}{\gamma LG}$$

The air-water interface will curve to form a meniscus when it comes in contact with solid surface. The meniscus angle with the solid surface, measured in the water, are lower than 90° (acute contact angles) when adhesive forces between solid walls and water molecules dominate the cohesive forces between the water molecules. On the other hand, obtuse contact angle are observed when cohesive forces dominate the adhesive forces. Acute contact angles are typical for interactions between pore water and soil solid surfaces. The curvatures of the menisci generate water pressures lower or higher than the air pressure depending on whether the contact angle is acute or obtuse angle, respectively.

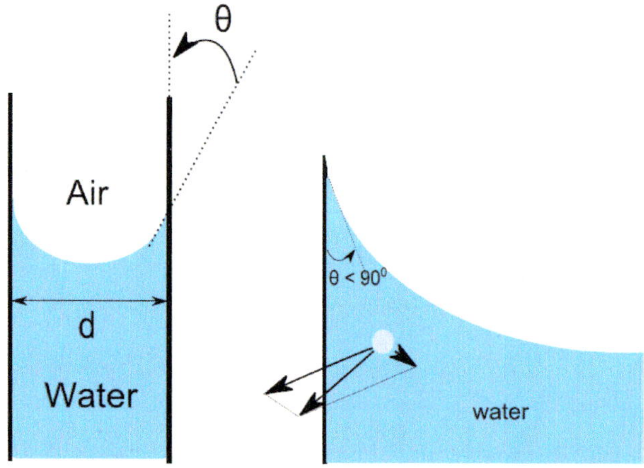

Contact angle formed when water is raised in capillary tube

(Source: https://archive.nptel.ac.in/courses/105/103/105103139/)

D. Capillary rise

The combination of adhesion forces, between water molecules and solid wall, and cohesion forces, between water molecules, causes water to rise in capillary tubes and soil pores above the free water level as shown in *Fig. below*. The adhesion forces cause the rise in capillary, and the cohesion brings all the water molecules together to follow the upward pull. The analyses by several researchers show that the wall of the capillary tube exerts an upward force on the water through the surface free energy difference (γ_{SG}- γ_{SW}), where γ_{SG} and γ_{SW} are the interfacial surface tension between solid-gas (*SG*) and solid-water (*SW*), respectively. As mentioned before, the concave curvature indicates the presence of pressure difference across the meniscus. The pressure below the meniscus will be smaller than the atmospheric pressure above the meniscus as shown in *Fig. below*. This is because the water in the capillary tube is suspended from the meniscus, which in turn is attached to the walls by hydrogen bonds. Therefore, the water is under tension, which is defined as "negative pressure". According to several other researchers, the pressure difference across the meniscus in the capillary is responsible for the rise of water in capillary tube.

Capillary rise is also explained in terms of the surface forces around the periphery of the meniscus. The straight-wall capillary due to upward force from the meniscus is balanced by the weight of the water column. The height of rise of water in the capillary tube, hc, can be expressed from this force equilibrium as:

$$h_c = \frac{4\gamma \cos \theta}{\rho g d}$$

where, γ is the surface tension (72.75 dynes cm^{-1} or 0.7275 dynes m^{-1}), ρ is the density of the water (1.0 Mg m^{-3}), g is the acceleration due to gravity (9.81 m s^{-2}), d is the diameter of the capillary tube (cm)

$$h_c = \frac{0.3}{d}$$

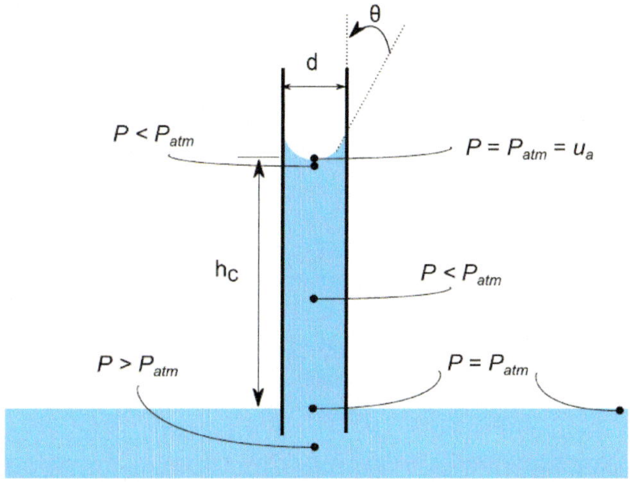

Capillary rise

(Source: https://archive.nptel.ac.in/courses/105/103/105103139/)

The concept of capillary rise is very important in unsaturated soil mechanics i.e. for understanding the natural moisture levels within the soil above the water table. Several models have been proposed by earlier researchers for predicting the ultimate height and rate of capillary rise in unsaturated soils based on the statistical variations in the pore geometries and hydraulic conductivities (fig below).

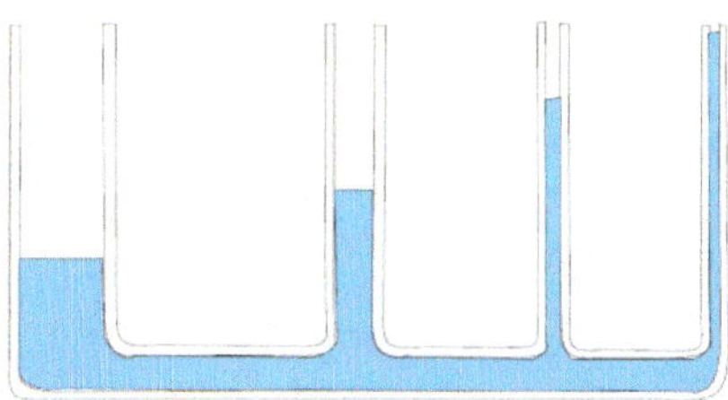

An illustration of capillary rise in tubes of different diameters

(Source: https://archive.nptel.ac.in/courses/105/103/105103139/)

E. Viscosity

Viscosity of water at 20°C is one centipoise (one hundredth of poise). Hydrogen bonding is the cause of high value of viscosity of water. The viscosity of water decreases with increase in temperature. So infiltration and percolation of water which are mainly influenced by the amount and type of clay and stability of aggregates are influenced by viscosity of water.

F. Vapour pressure

The vapour of the liquid at any temperature exerts pressure like a gas which is called the vapour pressure of the liquid at corresponding temperature. As the liquid molecules are in constant thermal motion, they colloid frequently and get sufficient momentum to escape out of liquid into atmosphere above it. At the same time, some of the molecules in the vapour phase may colloid with liquid and changes to liquid form. The pressure exerted by water vapour that is at equilibrium with free and pure water is called saturation vapour pressure. The vapour pressure that is at equilibrium with water depends upon the pressure, temperature ad solute content of water but does not depends upon the amount of water in liquid or gaseous form in the system. Because of the dipolar nature, vapour pressure of water is very low. The vapour pressure of ice and liquid water at 0°C is the same. The vapour pressure of boiling water is 1.0 atmosphere. Vapour pressure increases with increasing temperature because of increasing thermal motion of its molecules.

G. Dielectric constant (ε)

The dielectric constant (ε) of a solvent regards its capability to dissolve ionized solutes. A higher dielectric constant of the solvent means the higher ability of the solvent to dissolve salts. The dielectric constant exhibits the liquid polarity, and it is of primary importance for the evaluation of the solvent properties of water. Generally, the dielectric constant can be defined as the ratio of the absolute permittivity of a substance to the absolute permittivity of free space. At ambient condition, the dielectric constant of liquid water is around 78.4. At this condition, liquid water is poorly miscible with hydrocarbons and gases. Conversely, liquid water is a good solvent for the polar materials due to its high dielectric constant. With increasing pressure and temperature, the dielectric constant of liquid water substantially decreases.

SOIL WATER POTENTIAL CONCEPT

The concept of potential in soil was first applied by Buckingham (1907). The concept of soil water potential is a mean of stating the relative level of potential energy in different regions within the soil instead of absolute amount of potential energy contained in soil water. It represents the potential energy of the soil water in a particular location in the profile as compared to that of pure water at standard pressure and temperature in a standard referenced state (taken as reservoir of pure and free water at atmospheric pressure at the same temperature as the soil water).

The components of soil water potential due to difference in free energy resulting from gravitational, matric and osmotic forces are termed as gravitational potential (Ψg), matric potential (Ψm) and osmotic potential (Ψo), respectively. Each potential results from different forces and they are technically not additive. They act simultaneously to influence water behavior in soils.

Gravitational potential (Ψg): The components of soil water potential due to difference in free energy state resulting from gravitational forces are termed as gravitational potential. The gravitational force acting on the soil water is similar to that acting on any other body i.e. towards the centre of earth. Mathematically, gravitational potential is expressed as

$$\Psi g = gh$$

where g is the gravitational acceleration and h s the height of soil water above the reference level. Reference level is generally chosen within the soil profile or at its lower soil boundary so that the gravitational potential above the reference level is always positive.

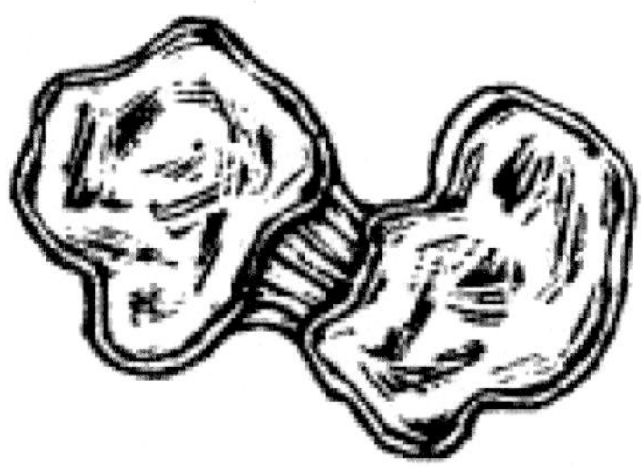

https://www.slideserve.com/nara/water-cycling-in-ecosystems

Gravitational potential is considered important for removing the excess water from the rooting zone following heavy precipitation or irrigation.

Matric (Ψm), or Osmotic potential (Ψo): Matric potential is the result of two phenomenon viz. adhesion (adsorption) and cohesion (capillarity). Matric potential may be defined as the amount of work done or energy required to pull a unit mass of water from reference level to the point in question.

The attraction of soil solids and exchangeable ions for water on the surfaces of soil particles and the attraction along with the surface tension of water in capillary pores caused loss of energy i.e. water held by the soil particles or in capillary pores is less free to move or react than the free water. Thus the work must be done to move that water. So, the adhesive and cohesive forces reduces the free energy of the soil water and due to this, matric potential can have zero (saturated soils) or negative values (unsaturated soils), the magnitude of which increases as the water is held more tightly in the soil. Matric suction has been defined as the negative of matric potential. Thus, matric suction is positive but matric potential is negative.

Osmotic potential (Ψo) arise from the presence of solutes in soil water which may be inorganic or organic compounds or both. These solutes reduce the free energy of the water because solute ions or molecules attract the dipolar water molecules. Water containing solutes are less able to do work than pure water. As work is required to remove salts from the water i.e. work is done against the solute suction, the osmotic potential will always be negative.

Potential may be expressed in terms of the height in cm of a unit water column whose weight just equal to the potential under consideration. The greater the height of the water in the column, the greater is the potential measured. Soil water potential also is expressed in terms of standard atmospheric pressure at a sea level which is 14.7 lb in $^{-2}$, 760 mm Hg or 1020 cm of water. One atmospheric pressure is approximately equals to one bar. One millibar is equal to 1/1000 bar. Ten bars is equal to one megapascal (Mpa) (SI units). Relationship among the three means of expressing the soil water potential is as follows:

Height of unit water of column (cm)	Soil water potential (bars)	Soil water potential (Mpa)
0	0	0
10.2	-0.01	-0.001
102.0	-0.1	-0.01
306.0	-0.3	-0.03
1020.0	-1.0	-0.1
15,300.0	-15.0	-1.5
31,700.0	-31.0	-3.1
102,000	-100.0	-10.0

CLASSIFICATION OF SOIL WATER

Physically, water may be classified based on the firmness with which it is held by the soil particles as -

1. **Hygroscopic water:** The water which is held by the soil particles mainly the soil colloids with a suction force of > 31 bars is known as *hygroscopic water*. This water mainly forms the first surface layer around the soil colloid. This water moves in the soils mainly in vapour form. This water is of no use as it is unavailable to the plants.

2. **Capillary water:** The water which is held by the soil particles with a suction force of 0.33 bars to 31 bars is known as capillary water. Here, the thickness of water layers around the soil particle consequently increases with the addition of some more water. The capillary pores (also known as micro pores) of the soil are filled with water and the suction of the soil water further decreases. The capillary water which is held with the suction force of 0.33 to 15 bars can only be available to the plants, above which water mainly moves in the soils in the vapour phase and is of limited use by the plants.

3. **Gravitational water:** Addition of more water in the soil filled the macro as well as micro pores of the soils. The water in the macro pores is loosely held with a suction force of less than 0.3 bar and will move downwards due to gravitational force, thus also called as *gravitational water or free water*. When macro and micro pores in the soil are completely filled with water, the soil is said to be at *saturation*. The gravitational and capillary water held in the soil is illustrated as below.

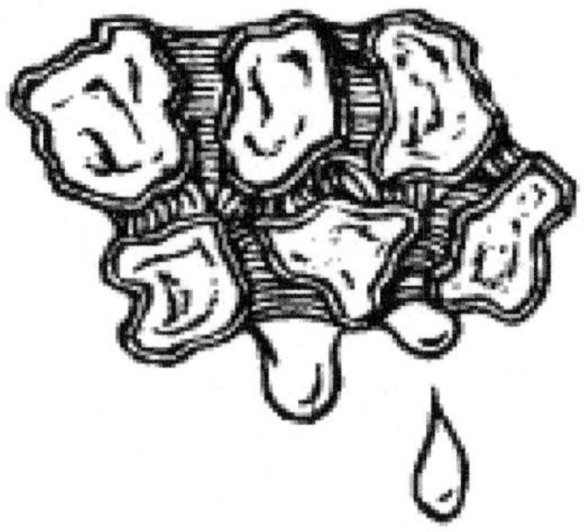

Gravitational water

The pore spaces are filled with water in excess of their capillary capacity, and the excess, or gravitational water drains downward

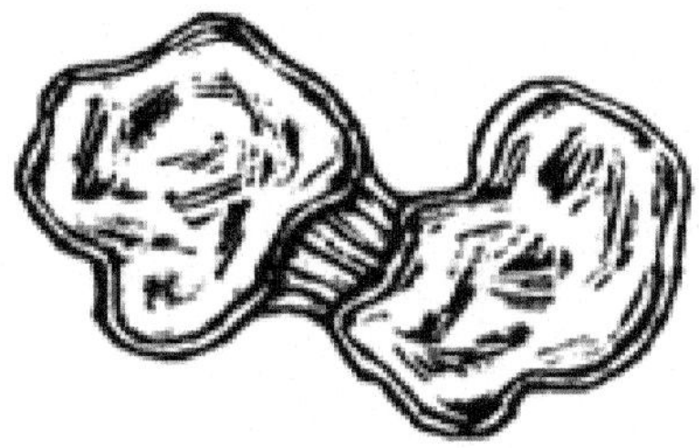

Capillary water

Capillary water is held in the pore spaces against the force of gravity

(Source: https://www.slideserve.com/nara/water-cycling-in-ecosystems)

Soil water retention: *Water is retained on the surface of soil due to forces of capillarity* (cohesion) *and adsorption caused by surface tension and surface attraction* (adhesion). Hydration of soil surface and exchangeable cations resulted in the adsorption of water. Water addition lead to increase in thickness of the water film adsorbed around the soil particles, then it gets into the soil pores. The amount of water retained in soil pores depends upon the surface tension force and the size of pores.

The water retention by the soil pores depends on their energy state with which it is held known as potential. In saturated soils, when all the pores (micro as well as macro) are filled with water, it is said to be at 0 potential i.e. water is not held by the soil with any force or it is free to move. This loosely held water, known as gravitational water, moves downwards in the soil due to gravity forces. Alternatively, if a light suction is applied to the saturated soils, no water may be released from the soil until the suction is increased to a certain critical value. This critical suction value is called the *air entry suction*.

The suction or the tension force with which the water is held by the soil particles increases with the decrease in the soil water after the drainage ceases. With further increase in the suction force or gradually over the air entry suction value (<0.33 bars), the macro pores which cannot retain water against applied suction get emptied first (suction 0.33 bars). As the suction increases beyond the air entry suction value, progressively micro pores will release water and at higher suction values, only very narrow pores retain water (suction > 0.33 bars). The thickness of the water films around the soil particles decreased with the increase in suction of soil water.

At equilibrium, the water held by the soil depends upon the sizes and volume of water filled pores and also on the matric suction. Here, the soil water is at sub-atmospheric pressure which is a negative pressure. In lieu of negative pressure, the term used is suction or tension.

Schofield (1935) introduced the scale of pF to express energy with which the water is held by the soil where F indicates the free energy of soil water. Therefore, pF may be defined as the logarithm of the height (cm) of a water column that is necessary to produce the desired suction. "*The relationship between the amount of water held by the soil at varying suction values is known as* ***soil moisture retention curve****/soil moisture release curve or soil moisture characteristic*".

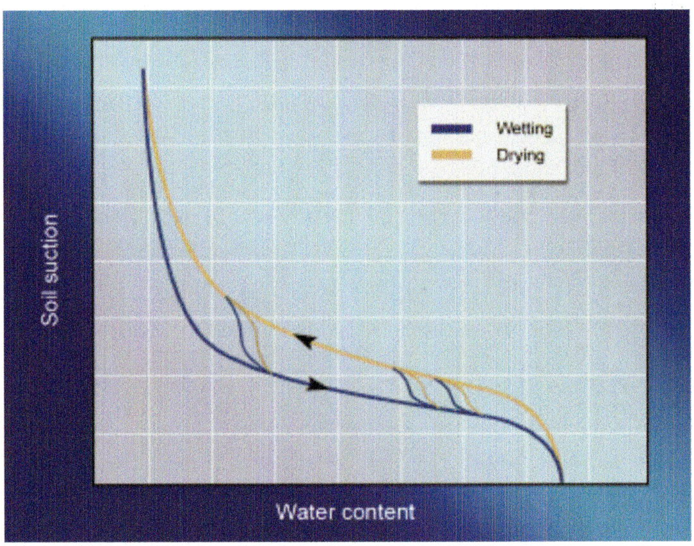

Factors affecting soil moisture retention are:

i) **Structure:** The amount of water retained at low suction (high matric potential) i.e. between 0 and 1 bar (100 kPa) strongly depends upon the capillary effect (total pore spaces) and pore size distribution of the soil which are highly affected by soil structure. Water retention at very high matric suction is not affected by soil structure because water retention at very high suction is related to surface adsorption on soil particles which depends upon specific surface of the soil. Water retention at low suction is high in well aggregated than compacted soil due to uniform pore size distribution.

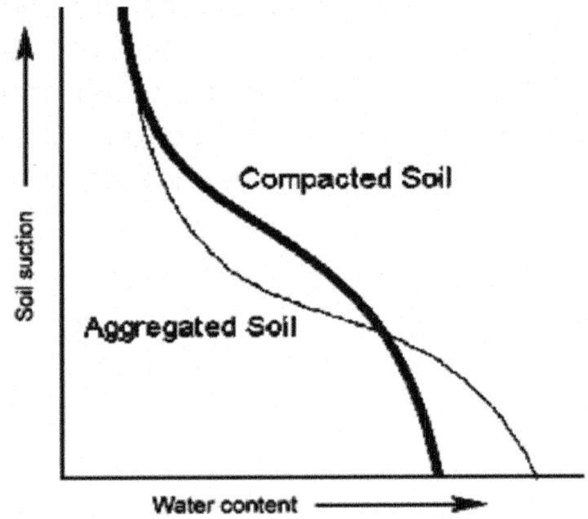

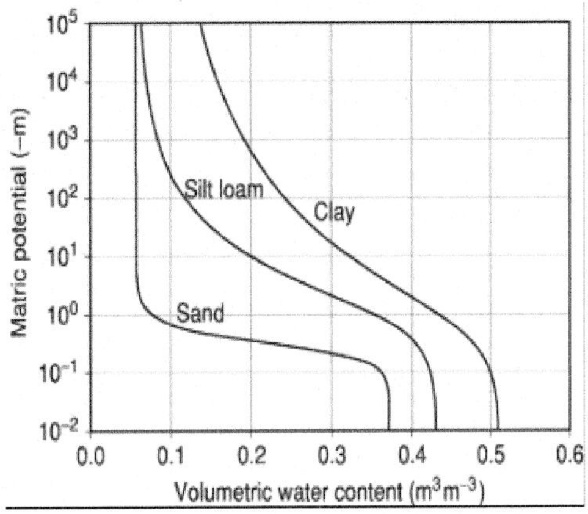

ii) **Texture:** The soil moisture retention curve is strongly influenced by the texture and specific area of the soil particles particularly at high suction due to adsorption of water by the soil particles. Soil water retention at particular matric potential increases as clay content increases. Therefore, suction will be significantly low in coarse textured soil as compared to fine textured soil at same volumetric water content. Therefore, soil texture influences the shape and range of the soil – moisture characteristic curve mainly in the high suction range.

iii) **Nature of clay minerals**: The nature of clay minerals also affects the water retention of soil at various suction/tension. Soils having montmorillonite as the dominant clay mineral (black soil) retain largest amount of water followed by the soils dominant in illite mineral (alluvial soils) and kaolinite mineral (laterite soils which have smallest amount of water). This variation in retention of water at varying matric suction is due to expansion and small specific surface area of the clay minerals dominated in black soils followed by Illite and Kaolinite.

iv) **Organic matter:** Organic matter has the greatest impact on retaining the water in the soil. The retention of water in soil is directly proportional to the content of organic matter in the soils i.e. more water will be retained in the organic soils. Organic matter helps in increasing the porosity of the soils by forming the stable aggregates which helps in retaining the water.

v) **Compaction:** Compaction decreases the total porosity of the soil by decreasing the macro porosity, thus causing decrease in water retention at low suction (0-1 bars) but amount of micro-porosity increases resulting in increase in water retention at intermediate matric suction. However, at high matric water retention is not affected by soil compaction because at very high matric suction, it depends on specific surface which is not related with compaction.

SOIL MOISTURE CONSTANTS

Water held by the soil at different suction values gives the knowledge of availability of water to the plants, water storage capability, percolation rate, water needed by the plants for its irrigation etc. Thus, the plot of soil water held at different tension forces known as soil moisture retention curve is a continuous function. Therefore, soil moisture retention curve is used to drive the different quantities of water retained by the soil corresponding to the specific water potential values which are of practical importance to the irrigation and drainage management of soil. Water held at specific water potential values are called *soil moisture constants.* These constants are used to characterize the soil by a single number to compare the retention capacity of different soils.

1. **Saturation or maximum water holding capacity:** When all the pores in the soil are completely filled with water, the soil is said to be at saturation. The amount of water held at saturation is called maximum water holding capacity of the soil. At saturation, water in the soil is at zero tension or suction. In practice, complete saturation may not be possible as some air always remains trapped in soil pores. Thus, for practical purposes, a soil is said to be saturated with water when the soil water is at zero tension or suction (suction is negative of the soil water potential) and majority of its pores are filled with water.

   The maximum water holding capacity of a soil is desirable in pot culture experiments as it provides a simple means of determining useful moisture level to be maintained for good plant growth. With medium textured soils in pots, good growth is obtained at moisture level corresponding to 50-70% of the maximum water holding capacity.

2. **Field capacity:** On draining the completely saturated soils, soil moisture decreased with the passage of time. After 2 to 3 days, the moisture in the surface layer attains a steady state. At this stage, the water retained in the soil is said to be at field capacity. Thus, field capacity is defined as *"the amount of water retained in the soil after the downward movement from a pre-saturated soil has materially ceased or practically zero".* The instrument used for determining the water retention by the soil at varying tension forces is *pressure plate apparatus*. The moisture with which it is held by the matric of the soil is affected by texture of the soil, amount and type of clay, organic matter content, depth of wetting, the presence of any impeding layers and evapo-transpiration rate of the soil. Generally, field capacity moisture content represents the moisture retained by the soil at 1/10 to 1/3 bar. In coarse textured soils, field capacity moisture content can be attained at 1/10 bars suction while it is 1/3 bars for medium or fine textured soils. The field capacity is about 4 % (by mass) in sandy soils, 45% in clayey soils and even upto 60% in some organic soils.

3. **Wilting point:** At wilting point moisture content transpiration rate of the plant increased and roots cannot be able to extract water to meet it, thus shows the symptoms of wilting as the plant loose its turgidity. At this point, plant cannot regain its turgidity after being kept it in a moist condition afterwards. Therefore, it is also called as permanent wilting point. Here, the water is held by the soil at the potential of -15 bars or with a suction value of 15 bars corresponds to a pF of 4.18. In laboratory, pressure plate apparatus is used for estimating the matric potential at permanent wilting point. The wilting point is affected by both the plant species and with varying plant growth stages. At wilting point, very few layers of water molecules are attached which is quiet thin, so at this stage the role of soil structure in water retention becomes negligible and only soil texture is responsible for it. Wilting point also got affected due to salts present in the soil. Generally, the roots of most of the plants are extended to sub-soils that may vary in moisture contents. Therefore, a plant can extract the water from the sub soils although it may be dry at the surface but has containing relatively higher moisture in lower depths.

4. **Hygroscopic coefficient:** *"The hygroscopic coefficient of soils is the amount of water held by the soil when it is in equilibrium with an atmosphere of 98% water vapour saturation or 98% relative humidity"*. The hygroscopic coefficient is the boundary between moist and dry soil and attains at soil moisture content of -31 bar. The quantity of water withheld by the soil at the hygroscopic coefficient depends upon the amount of clay minerals and its type, the nature of the exchangeable ions and salt concentration. Soils possessing high amounts of expanding type (2:1) of clay minerals and high organic matter have a high hygroscopic coefficient values.

5. **Oven dry soil:** Soil is considered as oven dry when it has reached equilibrium with the vapour pressure of an oven at 105°C. The pF value of oven dry soil is 7.0 or relative humidity is approximately zero. This condition is used as the basis for the calculation of soil moisture on gravimetric basis. The water is held by the oven dry soil at the potential of -10000 bars.

6. Other soil moisture terms:

1. ***Available water*:** The available water is that amount of water which the plant and its roots can easily extract from the soil for its growth. Thus, it corresponds to the water held by the soil at tension of 1/3 bars to 15 bars i.e. field capacity moisture content to permanent wilting point.

2. ***Air dry soil*:** The air dry soil represents the moisture content when soil moisture is in equilibrium with the atmospheric moisture content. It corresponds to -31 to -1000 bar depending upon the degree of saturation and temperature of the atmosphere. It is not an equipotential.

Soil moisture constants and their corresponding soil water potential are shown in Fig. below.

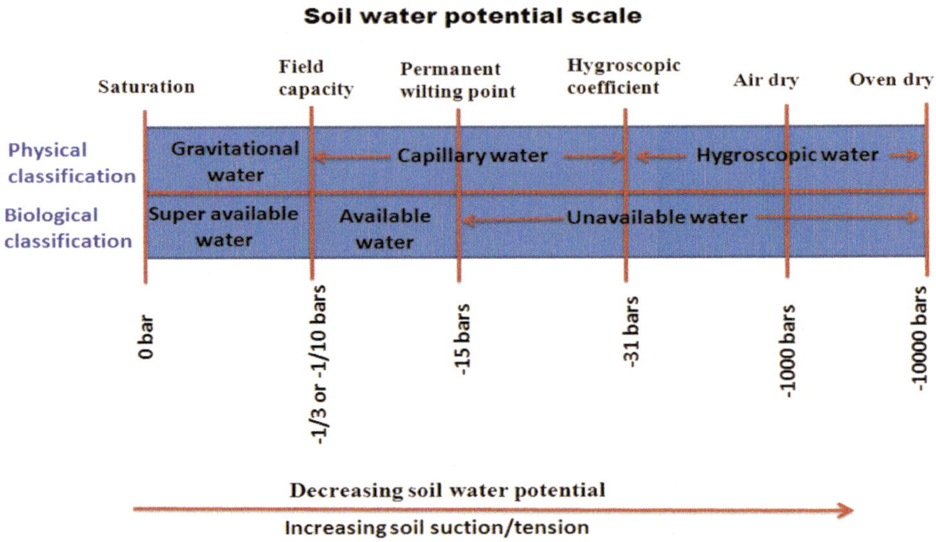

Soil moisture constants or soil water potential scale

METHODS OF MEASUREMENT OF SOIL MOISTURE

There are direct and indirect methods of measuring soil water. In direct methods, water is removed from the soil sample by evaporation. The amount of water lost is then calculated. In indirect methods, the property of the soil which is controlled by soil water content is measured. Directly, soil moisture is measured by the traditional thermo-gravimetric method and indirectly by several techniques based on the determination of electrical resistance, slowing down of fast neutrons, gamma-ray attenuation, electrical capacitance and backscattering coefficient as a function of water content.

Depending upon the purpose of soil-moisture determination, soil water can be expressed either in terms of its amount or tension. The amount of water in soil can be expressed either on mass or volume basis but, commonly, it is expressed as mass of water per unit of mass of oven dried soil. For example, if 1 kg of moist soil (soil + water) loses 0.25 kg of water upon oven drying then the mass of oven dried soil will be 0.75 kg. The soil moisture percentage on mass basis is calculated as:

$$\frac{0.25}{0.75} \times 100 = 33\%$$

On volume basis, the moisture content is expressed as the volume of water per unit volume of the soil. This measure gives a better picture of moisture available to roots in a given volume of soil.

1. **Thermo-gravimetric method:** In the thermo-gravimetric method, moist soil sample is weighed and then dried in an oven at 105°C to a constant weight, usually for 24 hours, in order to obtain the oven-dried weight of soil sample. The per cent soil moisture on mass basis is calculated by dividing the mass of water by the oven dried mass of soil sample and then multiplying by 100.

   This method is simplest, cheapest and most accurate, and widely used standard method. All other methods are standardized with this method. Demerit of this method is that it is destructive and repeated measurements cannot be taken at particular site at a given time. The method is arbitrary in the sense that some clay still contain appreciable amount of adsorbed water even at 105 °C. Further, some organic matter may oxidize and decompose at this temperature and hence, the loss of weight on drying may not be entirely due to evaporation of water. It is also laborious and time consuming as a period of 24 hrs or more is necessary before the results are available.

2. **Electrical resistance method:** It is one of the non-destructive, indirect and *in situ* methods of measuring soil water content. The method is based on the fact that electrical resistance of a soil volume depends on its water content. In this method, two platinum electrodes are embedded in a porous block of gypsum, nylon or fiberglass. The block is placed in the moist soil at desired depth and allowed to equilibrate. These blocks when embedded in soil tend to equilibrate with the soil moisture suction rather than with moisture content directly. The electrical resistance of the porous block is then measured by an electrical resistance meter based on the principle of the Wheatstone bridge. The electrical resistance of the soil decreases with increase in water content. The soil water content is obtained using a calibration curve of electrical resistance against known soil water content for the same block. The equilibration of these blocks may be affected by the texture, structure, soluble salt content and soil hysteresis. For these reasons and temperature sensitivity, resistance blocks are not applicable in soils which are light in texture, saline or sodic in nature. When using gypsum blocks, there is problem in achieving a desired accuracy as dissolution of gypsum with time destroys the gypsum block. But the method is easy and less time consuming, non-destructive and measurements can be repeated as and when required.

   The advantage of this method is that the porous block and wheat-stone bridge are relatively in-expensive. The disadvantage is that gypsum block deteriorates with time especially in acid soils, so each block has to be calibrated separately and repeatedly before reuse because of the changes internal porosity of the blocks.

3. **Nuclear radiations techniques:** The neutron scattering and gamma-ray attenuation are two nuclear techniques that can be used for non-destructive *in situ* measurement of soil water content.

a) ***Neutron scattering technique*:** In the neutron scattering technique, the instrument used is called *Neutron Moisture Meter*. It consists of americium and beryllium or radium and beryllium as the neutron source (9.25 x 10^7becquerel) and boron trifluoride (BF_3) gas as a detector. The instrument works on the principle that when fast moving neutrons are emitted radially into the soil, they undergo elastic collisions with hydrogen atoms having mass equal to the mass of neutron and get scattered. Their energy (kinetic) is reduced and speed is slowed down. (The neutrons will also colloid with atoms of other element such as Si, Al, etc., but the ability of hydrogen atom is greatest to slow down the neutrons because of same mass of neutron and H). The process is called *thermalisation* and the slowed neutrons are called *thermal neutrons*. The slowed neutrons are absorbed by BF_3 gas and counted with a scalar or rate meter or electronic microprocessor, which directly displays the moisture content. The density cloud of thermal neutrons formed around the source/probe is proportional to the concentration of hydrogen in the soil. Since water is the main source of hydrogen in soil, the density of slow neutron is proportional to the volume-fraction of water present in the soil.

The neutron moisture meter has to be calibrated for each soil. The probe which contains a source of fast neutron and a detector of slow neutron is lowered into an aluminum access tube inserted vertically into the soil to a desired depth. Aluminum access tube generally preferred as it is transparent to the neutron flux.

The main advantage of using neutron moisture meter is that it is an efficient and reliable technique for monitoring soil moisture in the field. It is rapid, less laborious, non-destructive and repeated measurements can be made at the same depth, and therefore, can handle large number of experiments. It is used for measuring moisture in deeper soil layers. The method is independent of temperature and pressure. The main disadvantages of the method are that it cannot be used for the measurement of soil moisture of surface soil since neutrons escape into the atmosphere which may be dangerous for the user. The instrument is costly also.

b) ***Gamma-ray attenuation technique*:** The gamma-ray attenuation technique of soil moisture content measurement depends on the fact that when a gamma-ray strikes to any substance, it loses part of its energy. The instrument used is called *Gamma-ray scanner or Gamma-ray apparatus*. It consists of two aluminum access tubes. In one of the tubes, radioactive isotopes ^{137}Cs (Caesium) or ^{241}Am (Americium) is used while other tube consists of thallium activated NaI crystal, which acts as a detector. When a narrow beam of gamma-radiation is passed through the soil, the rays are attenuated depending upon the thickness of soil mass, the bulk density and water content of the soil, following the principles of Beer' Law. If the thickness and the bulk density of soil are known, the water content of the soil can be determined. The gamma-ray apparatus is most useful in laboratory studies where moisture content of very small depth increments of soil is to be determined non-destructively. It is very rapid method but practical difficulties limit its use in the field. It is costly and there is a radiation hazard also.

A number of isotopes are used for soil water movement studies. The isotope tritium (^3H) is largely used for water movement, ground water recharge and pollution studies. In hydrological studies stable isotopes-deuterium (^2H) and oxygen (^{18}O) are used.

4. **Other methods:** Time domain reflectometry and microwave remote sensing are also in use now-a-days for the determination of soil moisture.

MOVEMENT OF SOIL WATER

The movement of water in soil takes place in three phases viz.

1. Solid,
2. Liquid and
3. Vapour phases.

Movement of water in the solid phase takes place when the soil gets frozen due to which the soil swells and thus movement of soil water in solid phase occurs as the entire soil mass as a whole.

Liquid water moves in the soils under saturated condition where all the pores are completely occupied by water. However, in a partially dry or wet soil, the movement occurs in liquid as well as in vapour phase. Flow of liquid water is of two types:

1. **Saturated flow:** Saturated flow occurs when essentially all the pores (micro as well as macro) are occupied by water and water is free to move with no tension force. This occurs in the soil after application of flood irrigation or during heavy rains, and in the groundwater zone.

2. **Unsaturated flow:** In unsaturated soils, water flow takes place when pores are partly filled with water and partly with air. The soil water under this condition is held with suction force i.e under tension.

The movement of water in vapour phase takes place by diffusion and mass flow:

a) ***Diffusion*:** Moving force is the gradient of partial pressure of any constituent member of variable gas mixture and it causes the movement of particular constituent member from a zone of high pressure to low pressure, the movement is called diffusion. Vapour pressure increases with increase in moisture content and temperature and it decreases with increase in salt concentration.

b) ***Mass flow***: In vapour phase, water moves along with other gases of the soil air as a whole due to the difference in total pressure gradient, the movement is called as mass flow. Mass flow can be caused by change in atmospheric pressure, soil temperature and pressure change cause by infiltration (compression) and percolation (evacuation). Mass flow of water vapour represents only a small portion of the entire water vapour movement in the soil.

The potential difference at any time between two points of the soils lead to the movement of soil water i.e. it moves from the point having higher potential to the point having lower potential.

Saturated water flow: ***Henry Darcy***, the French Engineer described the law of movement of water in the saturated soil called ***Darcy's Law*** which states that the flow rate (Q), which is the volume V flowing through the soil column per unit time (t), is directly proportional to the cross-sectional area (A) and the hydraulic head drop, Δh (Potential difference), and inversely proportional to the length of the soil column L (Fig. given below):

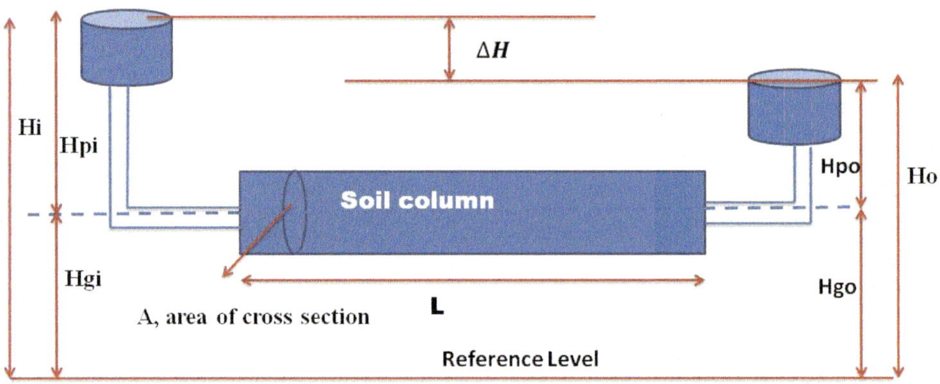

Figure: Flow of water in horizontal soil column

$$Q = \frac{V}{T} \alpha \frac{\Delta H}{L} \quad (2)$$

The hydraulic head drop (Δh) is given by:

$$\Delta H = H_i - H_o \quad (3)$$

$$H_i = H_{pi} + H_{gi} \text{ and } H_o = H_{po} + H_{go}$$

Where H_i & H_o represents the hydraulic head at the inlet and outlet of the soil boundary/column, respectively and H_{pi}, H_{gi} and H_{po} and H_{go} represents the pressure and gravity head at inflow and outflow end of the soil column, respectively.

In the absence of hydraulic head difference i.e. when $\Delta H = 0$, there will be no flow. The hydraulic head drop per unit distance in the direction of flow ($\Delta H/L$) is called as hydraulic gradient (*i*) i.e. it acts as driving force in the soils for the movement of water.

The rate of discharge of water per unit cross-sectional area, Q/A (i.e., volume of water *V* flowing in a unit cross-sectional area *A* per unit time *t*) is called specific discharge/flow velocity/flux density or simply water flux (v). Thus, the eq. 2 can be rewritten as:

$$v = \frac{Q}{A} = \frac{V}{At} \alpha \frac{\Delta H}{L}$$

$$v = K\frac{\Delta H}{L}$$

where v is also known as flux or Darcian velocity, and *K*, the proportionality constant is the hydraulic conductivity.

$$v = K\,i$$

This equation is known as Darcy's Law. Darcy's Law is valid when fluid is incompressible, medium is saturated, flow is in steady state and laminar, and temperature is constant. Under field conditions, the hydraulic head in most cases is such that the flow remains laminar.

Limitations of Darcy's law

The Darcy's law is valid in saturated water flow condition as long as flow is laminar and relationship between flux and hydraulic gradient is linear. This law is not valid at high flow velocities when the linearity of the flux versus hydraulic gradient relationship fails. Hence, Darcy's law is considered valid in all soils except in coarse sands and gravels where higher hydraulic gradients (more than unity) may result in non-laminar flow conditions. For saturated soils, flux remains linear with hydraulic gradient or flow of water is laminar as long as Reynold's number (R_n) is smaller than one.

Hydraulic conductivity: Hydraulic conductivityis the propertyof soil with which the soil allows the water to conduct through it. Hydraulic conductivity depends upon the properties of soil (size and distribution of soils pores) and fluids (density and fluidity).

Therefore, soil hydraulic conductivity, K (LT^{-1}) is expressed as:

$$K = k\,f$$

where k is *intrinsic permeability* (L^2) of the soil and depends upon the properties of soil only, and f is the*fluidity* (1/LT) of the permeating fluid (liquid/gas).

The fluidity is related directly to density and inversely to viscosity as follows:

$$f = \rho g / \eta$$

where ρ is density and η is viscosity of the fluid flowing through the soil.

Therefore, any change in fluidity caused by the change in temperature and soluble salts content results in change in saturated hydraulic conductivity.

Factors affecting the saturated hydraulic conductivity of the soil: Any factor which influences the size and distribution of the soil pores affects the soil hydraulic conductivity:

1. *Soil texture:* The hydraulic conductivity is highest in sandy soils and decreases with decrease in coarseness of the soil. Therefore, it is several times more in sandy soils than clayey soils under saturated conditions.
2. *Soil structure:* Hydraulic conductivity is higher in highly porous or well aggregated soil than in compacted or dense soil. Hydraulic conductivity is highest in soil having granular structure than the soils having platy structure.
3. *Organic matter:* Organic matter improves the hydraulic conductivity of fine textured soils due to formation of large amount of water stable aggregates.
4. *Fluidity:* Saturated hydraulic conductivity is also dependent on fluidity which depends upon concentration and composition of salts dissolved in water and viscosity of water.

Laboratory method of estimating saturated hydraulic conductivity

Hydraulic conductivity of the saturated soils can be estimated in the laboratory by taking the disturbed or undisturbed soil samples. Undisturbed soil samples are taken through core sampler while disturbed soil samples are air dried processed and repacked in a cylindrical container with a perforated bottom and open from top.

Hydraulic conductivity of the saturated soils can be estimated by

1. Constant head method
2. Falling head method

Constant head method is generally employed in those soils which have high permeability/conductivity i.e. more than 0.01cm min^{-1} while falling head method is used in relatively low permeable soils i.e. less than 0.01 cm min^{-1}.

a) **Constant head method**: In constant head method of estimating hydraulic conductivity, disturbed or undisturbed soil sample is taken in the cylindrical core, at the bottom of which filter paper was placed. The soil sample is saturated through capillary rise for overnight. A constant head of water is maintained from above the core sampler containing soil. The quantity of water flowing through a sample for a given hydraulic head is measured for a given time interval. Darcy's equation of water flow through a porous soil was used to estimate the hydraulic conductivity of the soil.

$$K = \frac{QL}{At\,(H + L)}$$

where, Q is the volume of water flowing through the soil of length L cm, having cross sectional area of A, t is the time, H+L is the hydraulic head in a soil column where the flow is vertically downwards, K being the hydraulic conductivity (cm hr^{-1})

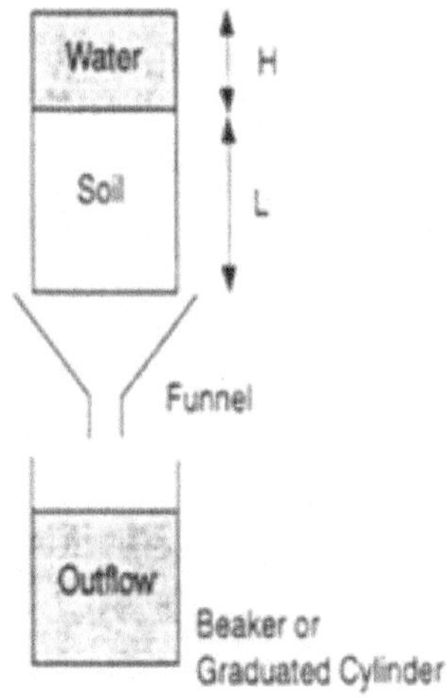

b) **Falling head method**: In low permeable soils or soils having low hydraulic conductivity eg. clayey soils, constant head method is not suitable for estimating hydraulic conductivity and the a capillary pipe is attached to the sample in which small quantity of water flowing through the soil is measured. The ratio of area of capillary pipe to the area of the soil can be adjusted to suit the hydraulic conductivity of the soil being tested.

In this method, a capillary pipe attached to the permeameter is filled with water and water is allowed to fall in the capillary from a head h_1 to a fall at h_2 in a particular time. The change in hydraulic head with time is recorded. Let the water fall from the initial head h_1 to the final head h_2 in a time t. If $-\frac{\partial h}{\partial t}$ is the small change in head w.r.t. small change in time, a is the cross sectional area of the capillary pipe, A and L are the area and length of soil column, respectively then the hydraulic conductivity K can be computed from Darcy's law as:

$$\frac{dQ}{dt} = KA\frac{h}{L}$$

$$dQ = KA\frac{h}{L}dt$$

This much quantity of water is measured in the capillary pipe is equal to dh, so

$$dQ = KA\frac{h}{L}dt = -a\,dh$$

$$\frac{AK}{aL}dt = -\frac{dh}{h}$$

Integrating both sides of the equation we get

$$\frac{AK}{aL}t = -\int_{h1}^{h2}\frac{dh}{h}$$

$$\frac{AK}{aL}t = -(\ln h_2 - \ln h_1) = \ln\left(\frac{h_1}{h_2}\right)$$

$$K = 2.3\frac{aL}{At}\log_{10}\frac{h_1}{h_2}$$

2. Unsaturated water flow

The water flow in unsaturated soils takes place when the soil water potential is generally less than -0.33 bars. In an unsaturated soils with uniform texture and structure, water moves along the gradient of matric potential from wetter to drier regions. The rate of flow increases as the water potential gradient (difference in potential per unit distance between the wetter and drier regions) increases and also as the size of the pores increases. Flow of water under unsaturated soils is more complicated and difficult to describe quantitatively due to variable soil wetness, suction, conductivity and also due to effect of hysteresis.

Water flow in saturated and unsaturated soils

1. In saturated soils, water flow takes place when all the micro as well as macro pores are filled with water i.e. water is not under tension. However, in unsaturated soils, macro pores are filled with air and finer micro pores are filled with water making the water under tension i.e. water held in the pores with the suction force. The free energy of the saturated soil is zero.

2. Moving force in case of saturated soil is the gradient of positive pressure potential and water flows from the zone of higher potential to the lower or lower suction to higher. In unsaturated soils water flows under the gradient of negative pressure potential termed as suction or tension.
3. In saturated soils, the most conductive pores are the macro pores while least is the micro pores. Thus at saturation, a sandy soil conducts water more rapidly than a clayey soil. However, in unsaturated soils clayey soils conducts water more rapidly than the sandy soils
4. In saturated soils, the hydraulic conductivity values are independent of the magnitude of water potential whereas in unsaturated soils, the hydraulic conductivity values varies with suction i.e. hydraulic conductivity values decreases with increase in suction values.
5. The actual flow path in saturated soils is less than that in unsaturated soils

Water infiltration: Infiltration is the entry of water into the soil due to sorption and vertical flow of water through the soil surface. The rate of water entry in the soil is termed as *infiltration rate* i.e. volume of water entering into the soil per unit area per unit time.

The infiltration of water determines/estimates the

1. Total amount of runoff/overland (runoff) flow,
2. Rate at which the irrigation to be applied
3. Textural and structural conditions of soil surface and
4. Transmission characteristics of sub-surface soil layers.

Infiltration rate is measured using double ring infiltrometer. Initially, the rate of infiltration is faster and subsequently reduces with time and attains a steady rate. The final steady rate is known as the *basic infiltration rate*. The total amount/depth of water that infiltrates the soil in a given time is called as *cumulative infiltration.*

Terminology related to infiltration

1. **Percolation:** Percolation represents the movement of water downward i.e. from one soil layer to another generally under saturated or nearly saturated condition. Infiltration is the prerequisite of the percolation. Percolation rate cannot exceed that of infiltration rate of the soil. Percolating water generally recharges the underground water table.. Percolating water carries away the dissolved plant nutrients specially Ca, Mg, S, K, N, P and other salts out of the zone of plant roots. This process of removal of soluble constituents with percolating water is called as leaching.

2. **Water intake:** The movement of irrigation water from the soil surface into and through the soil is called water intake.
3. **Overland flow:** It is the loss of water from the ground surface due to runoff when precipitation rate exceeds that of infiltration rate as a thin sheet of water. Low permeability favours the overland flow.
4. **Inter flow:** It is the lateral seepage of water in a relatively pervious soil above a less pervious layer.
5. **Seepage:** It is the vertically downwards and lateral movement of water into the soil from the source of supply like reservoir or irrigation channel.

Moisture distribution during infiltration in a homogenous soil profile

The initially dry, texturally uniform soil profile shows the different soil zones described as follows:

1. **Saturation zone**: Upon ponding, the surface soil is completely saturated to a depth of few millimeters and this is known as saturated zone.
2. **Transition zone:** Just beneath the saturation zone, there is a rapid decrease of water content with depth and this is known as transition zone.
3. **Transmission zone:** This zone is just below th transition zone having uniform wetness which is slightly less than complete saturation i.e. wetness is near to saturation. The transmission zone is the lengthening zone and the thickness of this zone increases with increase of infiltration time.
4. **Wetting zone:** Below the transmission zone, the water content sharply deceases with depth and this zone is termed as wetting zone.
5. **Wetting front:** This forms a sharp boundary between wet and dry soil and beyond this there is no penetration of liquid water.

Factors affecting infiltration rate:

1. *Pore size*: Infiltration capacity usually increases as the size of pores increases.
2. *Initial moisture content*: Drier the soil more will be the infiltration rate.
3. *Surface condition*: The highest infiltration capacity is normally found in protected forest land > pasture > crop land > fallow.
4. *Time*: The infiltration capacity of a soil is highest at the beginning and then decreases until constant rate is obtained i.e. infiltration rate decreases with increase in time.

Infiltration rate with time is not similar in different soils. The change in infiltration rate with time is least in the sandy, gravelly or rocky soil. In clayey soils, the greatest changes occur as these soils crack on drying and thus, allow unlimited infiltration at the beginning. Once the cracks are filled with water, swelling of clay takes place and cracks are closed, therefore, infiltration practically ceases. In addition to texture, the infiltration rate increases under vegetation as compared to bare soil.

The time changes of cumulative infiltration (I) and infiltration rate (i) are described by the equations given by Philip (1957)

$$I = s\, t^{1/2} + Kt$$

$$i = \frac{1}{2} s\, t^{-1/2} + K$$

where, s is the sorptivity function and K is the hydraulic conductivity of the soil corresponding to the water content prevailing in the transmission zone that occurs below the top saturated zone.

Water percolation: Percolation is the downward movement of water through the soil profile. It is the same as drainage. The downward movement of water below the root zone depth is called *deep percolation/drainage*. It is loss of water which is more in coarser than in finer soils. Pressurized/micro irrigation systems (sprinkler and drip) are helpful in checking deep percolation loss.

Percolation is very important in soil development and land management. It removes the high salt content through leaching but removes valuable plant nutrients also. In coarser textured soils, the percolation capacity is greater as compare to the finer textured soils.

Chapter 13

Soil Air and Aeration

The soil air and aeration is very important for optimum plant growth. The soil air consists of nitrogen, oxygen, carbon dioxide, inert gases and water vapour similar to that of atmospheric air. The atmospheric air, by volume, is composed of approximately 79.0% N_2 and inert gases, 20.79% O_2 and 0.03% CO_2. The composition of soil air is more or less the same to that of the atmospheric air except CO_2 content. The composition of air in surface soils with optimum porosity constitutes 79.1% N_2 and inert gases, 20.6% O_2 and 0.25% CO_2. The CO_2 content of the soil air is approximately 10 times more than atmospheric CO_2 as because the respiration by soil fauna and flora will consume O_2 and evolves CO_2. However, in reduced water logged soils, soil air may contain a substantial amount of methane, hydrogen sulfide, and ammonia as these gases are formed in the process of decomposition of organic matter under reduced or waterlogged conditions.

The air which is the mixture of gases envelops the earth and extends up to a considerable height from the surface of the earth. This envelop is known as an atmosphere. The atmosphere is divided in to a number of horizontal layers parallel to the surface of the earth, each layer bears the weight of all other layers above it and is thus subjected to a pressure due to its weight. This pressure is called atmospheric pressure and is greatest at the surface of the earth and decreases with an increase in height above the earth's surface.

The standard atmospheric pressure may be defined "as the pressure exerted by a vertical column of zero degree cold mercury, 76 cm in height and density 13.595 g cm^{-3} subjected to the gravitational acceleration of 980.665 cm s^{-2}". Therefore, one normal or standard atmospheric pressure is equal to 76 x 13.595 x 980.665 = 1,013242 dynes cm^{-2} and a pressure of 1.0 million dynes cm^{-2} is called a bar.

"1 bar = 10^6 dynes cm^{-2}"

"1 bar = 100 milli bar"

Thus 1N or "Standard atmospheric pressure = 1013.24 millibar"

"1 atmospheric pressure = 1.013 bars"

IMPORTANCE OF SOIL AIR

1. Oxygen, the most important component of soil air, required during respiration by flora and fauna present in the soil.
2. The CO_2 of the soil air helps in making the nutrients available to the plants by essentially dissolving them in soil solution.
3. Nitrogen, in gaseous form, is present in the soil air in large quantity which through the microorganisms can be utilized by plants for their growth.
4. Water vapour serves as a source of water at the time of complete dryness in the soil and thus can save the plant due to desiccation of soil.

The oxygen should be regulated in the soil which is required for the respiration of root of plants and microbes. The vigorous growth of the plant and higher plant population leads to increase the oxygen requirement of the soil. Minimum 10% concentration of soil air is needed for normal growth of the plant. However, excess of oxygen also is not beneficial as it may lead to increase the oxidation of organic matter leading to drying of the soil quickly, which may retard the growth of plant roots. The reduction in oxygen concentration of the soil is highly fatal to the plant roots than the escalated concentration of CO_2 in the soil which should be in a permissible limit (i.e. < 20% CO_2).

Coarse textured and high organic soils falling under the high temperature zone will lead to excessive aeration, especially during the dry spell. Thus, adequate aeration of the soil is very important for the proper growth of plant. *Adequate aeration* can be expressed *"as a soil condition in where the oxygen diffusion rate (ODR) is at least 30 x 10^{-8} g cm^{-2} min^{-1} and O_2 concentration of the soil air is at least 10% in the root zone"*.

SOIL AERATION

Soil aeration is defined *"as the process of exchange of O_2 and CO_2 between the soil and the atmospheric air"*. It helps in replenishing the O_2 consumed in respiration by plant and microorganisms, which in-turn prevents the higher buildup of CO_2 (Fig. below). Aeration porosity expressed to describe aeration status of soil is defined *"as the volume of pore space filled with air when soil is under a tension of 50 cm of water. At this tension, soil pores of sizes greater than 0.06 mm in diameter are filled with air"*.

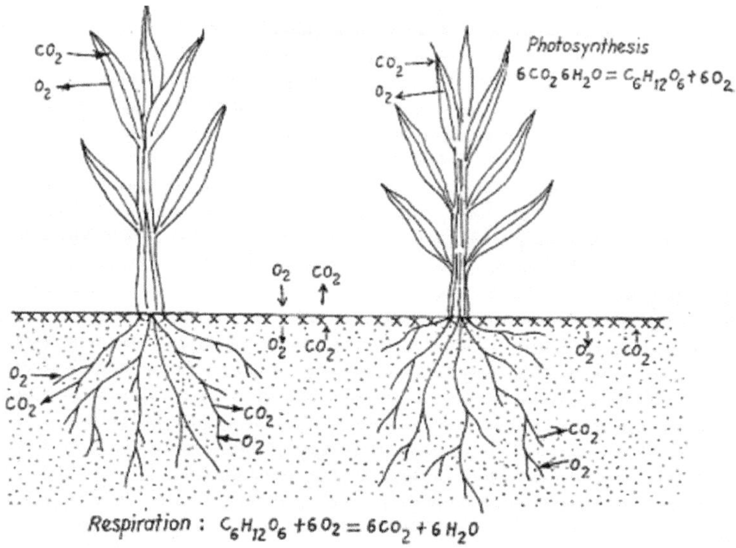

Fig. Aeration status of the soil due to respiration by roots and microorgansims

(Source: Text book of Soil Physics- A.K. Saha Edn)

FACTORS AFFECTING THE COMPOSITION OF SOIL AIR

The consumption of O_2 and production of CO_2 due to activities and respiration by plant roots and microorganisms rapidly change the composition of soil air. This change in composition of soil air will be renewed by atmospheric air through exchange of gases from soil air to atmospheric air and vice-versa. Following are the factors that may affect the aeration status of soil.

1. **Soil physical properties:** The porosity of soil air may affect the composition of the soil. Soils with higher macro porosity may get emptied immediately after heavy rains or irrigation and filled with air, however micro pores may takes few days to change the composition of soil air after these are filled with water completely due to high porosity. When the water from the soil drains or evaporates from the soil surface, air enters into the soil pores (primarily macro-pores) and volume of soil air increases. The volume of macropores affects the total air and gaseous exchange. Fine textured soils have less aeration due to more nos. of micro pores as compared to coarse textured sandy soils in which macropores dominate. Similarly, clayey soils have more organic matter content which may hold water for longer period of time, thus the composition of soil air is less owing to restricted aeration.

   Also, the aeration decreases with depth due to increase in compaction and bulk density of the sub-soils. The increased moisture content and reduced porosity at sub soils is another factor for decrease in O_2 concentration of the sub soils.

2. **Respiration:** Soil microorganism consumes O_2 and release CO_2 during respiration which may highly affect the composition of soil air. The respiration may also be affected due to change in temperature of the soil, moisture content of the soil and type of tissues involved in respiration.
3. **Soil organic matter**: The composition of soil air is highly influenced by the presence or absence of organic matter in the soils. Decomposition of organic matter due to microorganisms will consume O_2 content of the soil and produce CO_2. Thus, addition of organic materials in the soils highly influences the composition of soil air.
4. **Biological activity:** Higher biological activity in soils may highly influence the composition of soil air due to change in the composition of CO_2 and O_2 caused by respiration. High microorganism activity, especially nitrogen fixing bacteria, may lead to higher fixation of atmospheric nitrogen in the soil.
5. **Cropping:** Higher plant population and higher rooting density may influence the soil aeration by reducing the O_2 content due to higher respiration and increased in the concentration of CO_2.
6. **Tillage:** Tillage activities may also influence the composition of soil air. Tillage will loosen the soil, thus will increase the gaseous exchange of soil air with atmospheric air. On the other hand reduced tillage and no tillage will decrease the exchange of gases from the soil to atmosphere. Puddling of soils will decrease the pore spaces and thus resulted in poor aeration of the soil.
7. **Season:** Seasonal variation may affect the composition of the soil air primarily due to change in climate. Higher CO_2 concentration observed during the rainy season was due to higher microbial activity than O_2 content. However, during the dry season in summer, O_2 content of the soil air is higher than that of CO_2 content due to lowering of microbial activity. But in warm and humid season, due to more microbial activity, the CO_2 concentration of the soil far reaches that of the O_2 concentration.

Concisely, the concentration of CO_2 of soil air is higher in the

- Summer than in winter.
- Manured, limed, fertilized and vegetated soils because of greater root and microbe activity.
- Moist soils than that in dry soil because water may occupy the pores thus restricting the diffusion of soil air

- Clayey soils than in sandy soils because of minimal diffusion rates because of the dominance of micro pores which holds water for longer period of time..
- Highly compacted soils with high bulk density (i.e. for puddled soils) than in the soil with crumb or granular structure because of the difference in diffusion rates; and
- Sub-soil than in the top-soil as the diffusion of gases takes place directly from top- soil to atmosphere and vice versa while in sub-soil, this exchange takes place from subsoil to top-soil and vice versa.

Processes of gaseous exchange

The gaseous exchange takes place through the process of mass flow and diffusion.

1. **Mass flow:** It may be defined as the "movement of soil air as a whole from the zone of higher pressure to the lower due to total pressure gradient of air". Thus the moving force in case of mass flow is the gradient of positive pressure potential and the air will move from higher to lower pressure potential. This flow of air may occur from one point of the soil to another, or from atmosphere into soil or vice-versa. The difference in pressure of the soil air may be the result of difference in temperature, pressure, air replacement due to rainfall or irrigation.

i) ***Soil temperature:*** Charle's Law states that, "air pressure is directly proportional to the air temperature at a constant volume". Thus the temperature gradient will develop pressure gradient due to which the soil air will move from higher gradient to lower. Temperature difference may arise within the soil layers or between the soil and the atmosphere. The heat move in the soil up to the depth of 30 cm, thus the air movement is also restricted up to this depth and below which there is uniform composition of soil air.

ii) ***Pressure:*** Boyle's Law states that, "the volume of air is inversely proportional to the pressure at a constant temperature". When there is an increase in the atmospheric air pressure, the volume of soil air decreases resulting in movement of atmospheric air into the soil. Similarly, the decrease in the atmospheric air pressure will increase the volume of soil air, leading to the movement of soil air into the atmosphere.

iii) ***Rainfall and irrigation:*** Flooding the soils due to heavy rainfall & irrigation will occupy the soil pores with water and thus displaces the soil air of the pores. Partial pressure of CO_2 will increase in the surface soils and oxygen in the subsurface. Thus, the increase in O_2 concentration at lower soil depths will increase and decrease at surface.

iv) ***Wind:*** Pressure and suction effect of high wind causes the exchange of gases between soil and atmosphere but the effect is restricted to surface soil only.

2. **Diffusion**: Most of the gases in the soil move by the process called diffusion. Diffusion is the most important process of movement of gas in the soil. According to the kinetic theory of gases, "the molecules of gases are in a state of random motion in all directions". When two gases are put together, they readily get mixed up. This is known as diffusion. Moving force in diffusion is the gradient of partial pressure of any constituent member of variable gas mixture such as air and it causes the movement of that particular constituent member from the zone of higher pressure to lower although the soil air as a whole remains the same.

   For instance, the soil air contains a higher amount of CO_2 and lower amount of O_2than the atmospheric air. Thus, the partial pressure of CO_2 is higher in the soil air than in the atmospheric air, and the partial pressure of O_2 is higher in the atmospheric air than in the soil air even though the total pressure in the atmosphere and soil may be same. This results in the diffusion of CO_2 from soil into the atmosphere and O_2 from atmosphere in to the soil to maintain the equilibrium. Diffusion increases with increase in temperature.

Fick's Law of diffusion

Adolf Fick (1855) formulated this law which states that the rate of diffusion of any gas molecule is directly proportional to the change in its concentration per unit change in the distance, i.e. concentration gradient of that gas.

Mathematically it is expressed as

$$q_i = -D\frac{dc}{dx} \qquad (1)$$

where q_i is the flux of gas which is defined as the amount of gas diffusing across a unit cross sectional area in a unit time,

D is the diffusion coefficient or diffusivity,

dc/dx is the small rate of change in concentration with a small distance, dx i.e. concentration gradient.

The equation for ideal gases is:

$$PV = nRT \qquad (2)$$

where P is the partial pressure of gas, V is the volume of gas, n is the number of molecules of gas, R is the gas constant (8.3 joules) and T is the absolute temperature.

Therefore, $\frac{n}{V} = c = \frac{P}{RT}$ where c = concentration

Now from the equation (1) we get

$$q = -D\frac{d}{dx}\left(\frac{P}{RT}\right) = -\frac{D}{RT}\frac{dP}{dx} = -D'\frac{dP}{dx}$$

where D' is the diffusivity when driving force expressed in terms of partial pressure gradient.

The D' diffusion coefficient of a gas in soil air (D) is smaller than that of atmospheric air (D') due to the limited and tortuous nature of the soil pores.

Penman (1940) related these two as: $\frac{D}{D'} = 0.66\ f_a$ where f_a is air filled porosity.

According to Fick's II Law, the rate of change of concentration at a distance from origin is directly proportional to the rate of change of concentration gradient.

$$\frac{dc}{dt} \propto \frac{\partial}{\partial x}\left(\frac{\partial c}{\partial x}\right) \text{ or } \frac{dc}{dt} = D\frac{\partial}{\partial x}\left(\frac{\partial c}{\partial x}\right) \text{ or } D = \frac{\partial^2 c}{\partial x^2}$$

The coefficient of diffusion of gases (D) depends on the structure and moisture content of the soil. It is proportional to the air filled porosity. The rate of diffusion increases with temperature. There is a difference in diffusion of gases from the dry and moist soils having coarse and fine texture. In fine textured dry soils, the diffusion of gases is higher due to higher porosity of fine texture soils which is filled with air than that of dry coarse textured soils in which the air filled porosity is less. However, in a fine textured moist soil, the air filled porosity is less as most of the pores are filled with water than the coarse-textured, thus diffusion is more in coarse textured moist soils.

CHARACTERIZATION OF SOIL AERATION STATUS

Following steps/methods can be adopted to characterize the aeration status of the soils.

1. **Determining the composition of soil air:** The composition of soil air particularly of O_2 and CO_2 can be done using chemical or gas analyzer. Gas chromatography can also be used to characterize the composition of the soil air.

2. **Air-filled porosity:** The air filled porosity (f_a) of a soil can be determined by finding total pore space (f) and water content on volume basis (q_v) by the equation:

$$f_a = f - q_v$$

3. **Amount of carbon dioxide released:** The roots and microbial activities will estimate the amount of CO_2 released from the soil which in turn will help in characterizing the aeration of the soil.

4. **Air-permeability:** It indicates the ease with which the soil pores transmits air through it when the pressure gradient develops. The permeability of the soil to air depends upon the the ratio of pores filled with air, their size (micro or macro) and the continuity of pores to each other. The instrument used for determining the air permeability in field and laboratory is *Air permeameter*.

5. **Diffusion coefficient:** The exchange of gases between the soil and the atmosphere takes place mainly through the process of diffusion. Thus, the diffusivity coefficient is used to characterize the soil for its aeration status.

6. **Redox potential:** Oxidation-reduction potential or redox potential (E_h) is an important chemical characteristics of soil related to soil aeration. It indicates the tendency of a system to reduce or oxidize elements of the compounds.

The oxidized state of the element such as ferric (Fe^{3+}), manganic (Mn^{4+}), nitrate (NO_3^-), and sulphate (SO_4^{2-}) exist in the soils characterized by high aeration status while in a poorly aerated soils these elements exists in its reduced state viz. ferrous iron (Fe^{2+}), manganous manganese (Mn^{2+}), oxides of nitrogen (NO_x) and sulphide (S^{2-}).

The E_h provides a measure of the tendency of a system to reduce or oxidize substances. The values of E_h can be interpreted up to a certain extent in terms of the ratios of certain oxidizing and reducing organic and inorganic materials present in the soil system. The value ofE_h is positive and may be as high as +0.5 volt under strong oxidizing conditions. It is low and even negative when elements are present in reduced forms. The E_h values have the advantages of providing a measure of intensity of the reduction under reduced conditions. It is, therefore, useful in characterizing the soil aeration in the range below which the diffusion of oxygen is zero, for example, the rice fields where soils remain saturated or submerged under water.

The E_h values of certain elements at which their oxidized and reduced forms occur are given as below:

The oxidized and reduced forms of certain elements and their redox potential (E_h) at which the exchange in their forms occur.

Element		E_h at which exchange in form occurs (V)
Oxidized	Reduced	
O_2	H_2O	0.38 — 0.32
NO_3^-	N_2	0.28 — 0.22
Mn^{4+}	Mn^{2+}	0.28 — 0.22
Fe^{3+}	Fe^{2+}	0.18 — 0.15
SO_4^{2-}	S^{2-}	-0.12 — 0.18
CO_2	CH_4	-0.20 — 0.28

7. **Oxygen diffusion rate (ODR):** The ODR is defined "as the rate at which O_2 can be replenished when it is used by respiring plant roots or by microorganisms". It can be measured through ODR meter. This instrument contains a small platinum electrode and a standard calomel electrode. These electrodes are allowed to insert in the soil up to the required depth. The calomel electrode is connected to the galvanometer which is connected to the positive end of an external current source and the negative end is connected to the platinum electrode. The output electric current is related to the rate of oxygen flux at the electrode surface and given by the equation:

$$i'10^{-6} = nFA \quad (1)$$

where

"i - electric current (microamperes),

n - number of electron required for reducing one molecule of oxygen which is 4,

F -Faraday's constant,

A - surface area (cm^2) of platinum electrode, and

φ - flux or ODR to the electrode surface in number of molecules of oxygen per second per

cm^{2}".

The ODR is calculated by equation (1) as:

$$ODR = \frac{i \times 60 \times 32 \times 10^{-6}}{4 \times 96500 \times A} = \frac{0.497 \times 10^{-8} \times i}{A} g\,cm^{-2}min^{-1}$$

The value of ODR is critical for the optimum growth of the plants. Roots of the plants ceases at ODR fall to a value of 20×10^{-8} g cm^{-2} min^{-1}. The optimum ODR at which the plant growth is normal is about 30-40 x 10^{-8} g cm^{-2} min^{-1}.

The requirement of ODR of some crops is given as below:

Crop	ODR (10^{-8} g cm^{-2} min^{-1})
Maize	30.0
Soybean	28.4
Wheat (unirrigated)	26.9
Wheat (irrigated)	26.3
Peas (Unirrigated)	32.9
Peas (irrigated)	39.5
Toria	22.9

Sub soils have lower ODR than the surface soils due to decreased porosity and the high bulk density of sub soils. With an increase in moisture content of the soil, ODR decreases due to decrease in the aeration porosity.

Effects of soil aeration on plant growth

The plants and microorganisms and their activities have direct and indirect effect on the aeration status of the soils. The extent of soil aeration determines the forms of several inorganic elements available to the plants, reactions occurring in the soil, decomposition of organic matter, existence of symbiotic and non-symbiotic nitrogen fixing microorganisms and in turn, soil properties and plant growth.

The ill effects caused due to poor aeration are:

i) Plant roots require O_2 for their respiration, without which the growth may be retarded or may show abnormal growth. In root crops like sugar beet, carrot, turnip etc., there may be de shaping of the roots due to prolonged water logging.

ii) Nutrient movement and its uptake by the roots take place under oxygen-rich environment which becomes negligible in poorly aerated soils.

iii) There may be an increase in the incidence of disease in the absence of oxygen.

iv) Certain bacteria which help in nitrogen fixation, nitrification processes may get affected due to lack of sufficient O_2 as they require oxygen for their respiration.

v) Decomposition of organic matter also slows down due to lack of sufficient O_2. In the complete absence of O_2, anaerobic bacteria may decompose soil organic matter with the release of few toxic substances as H_2S, methane gas, toxic organic acids, etc., which are injurious for plant growth.

vi) In well oxidized soils, the oxide forms of most of the elements are present, e.g. nitrate-N. These are desirable for most of the crops. Anaerobic conditions favour the formation of some toxic inorganic compounds which are injurious to the plants.

Management of soil aeration

Any management practice that may increase or decrease the exchange of gases in to and through the soils is useful for growth of plant. Mostly the movement of gases in the soils takes place through the process of diffusion. Thus, management should be such that may increase the diffusion rate of the soil for restoring proper aeration. The porosity, especially the air filled porosity, determines the diffusion of the gas into the soil. The size of pores has a very little effect on the aeration status of the soil. Following are some of the management practices that may improve the aeration status of the soil.

1. **Soil structure:** Soil structure is very important for aeration of the soil. The improved soil structure has a higher amount of pores filled with air. The structure of the soil can be attained by using the amendments in the form of manure and compost on a regularly basis. The surface and sub-surface crust may restrict the movement of gases in to and through the soil. Also, the fine pores of the crust may retain water for a longer period of time, thus will prevent diffusion of gases. Thus, it is recommended to avoid formation of the crust for the proper growth of the roots.
2. **Drainage:** Optimum plant growth can be obtained from well-drained soils due to the continuous exchange of gases from it. In poorly drained soils, the aeration status of the soil is poor due to the limited gaseous exchange. So the proper drainage system should be installed for the gaseous exchange to take place. The improved drainage will decrease the water content of the pores which then get filled with air. .
3. **Cultivation:** The cultivation of soil using tillage implements will increase the exchange of gases due to loosening of soil particles. Deep cultivation of poorly drained soils is recommended for the optimum gaseous exchange. Cultivation followed by light rains will develop the surface crust on sandy loam soils and thus restrict the aeration of soil. This type of crust can be managed by breaking the surface crust with the tined hoe. The deep crust can be managed by growing deep-rooted crops and applying vertical mulch.
4. **Temperature:** Temperature has both positive and negative effect on the aeration status of the soil. The pressure/concentration of O_2 increases with the increase in temperature but simultaneously increasing the temperature will also increase the microorganism activities with the result of production of CO_2 due to higher respiration in soils.

5. **Plant adaptations:** Plant roots require oxygen for respiration, but some plants are adapted to grow under restricted aeration condition. Most of the plants are adopted to survive under oxidized conditions with only few, that can respire under O_2 free environment. Thus, the selection of the crop is to be done according to the aeration status of the soil. For eg. Rice can be grown in waterlogged condition probably because of its ability to respire O_2 from the leaves and cortex cells of the plant.
6. **Regulation of root respiration:** Regulation of respiration of roots is also equally important for the gaseous transformation in the soil. It can be done through adopting cultural practices, the decomposition rate of residues, fertilization, etc. Surface soils should be manured properly so as to keep the surface free from any crust formation etc.

Chapter 14

Soil Temperature

Temperature is very significant growth factors for crops and its regulation in soil helps in carrying out various processes that may affect the movement of air, water and hence the nutrients. Soil temperature governs evaporation, aeration, and rate and type of reactions viz. chemical and physical occurring in the soil. Soil temperature influence most of the biological processes occurring in the soil viz. germination of seed, its emergence and growth, development of root and the other microbial activities. Soil temperature in relation to plant growth is better understood by the following the terms given below:

1. **Temperature & heat:** Heat, considered as a form of energy, is being measured by temperature. Thus, heat energy is the energy of motion called the kinetic energy of the molecules of a substance due to their random motion or vibration. Heat can be transformed in different forms of energy. For eg. in a steam engine, heat produced by combustion of coal transforms water into steam and this steam sets the engine to run.

   The C.G.S. unit of heat energy is *calorie* and S.I. unit is *joules.*

   Calorie is defined "as the amount of heat needed to raise the temperature of one gram of pure water from 15°C to16°C".

   The temperature is expressed/measured in degree-Centrigrate (°C) or degree-Farenheit (°F). It can also be measured in degree-Reaumur (°R).

   The relationship between the three units of temperature is given as

$$\frac{C}{100} = \frac{F-32}{180} = \frac{R}{80}$$

$$\frac{C}{5} = \frac{F-32}{9} = \frac{R}{4}$$

Absolute temperature can be measured on Kelvin which is equal to 273.18 plus temperature in °C.

Latent Heat: It is expressed "as the quantity of heat required to change the state of matter without changing the temperature". If the matter changes from solid to liquid state it is called as *latent heat of fusion* and from liquid to vapour state, it is called as *latent heat of vapourization.*

For eg. If the latent heat of fusion of ice is 80 calories, means 80 calories of heat is required to change 1 gm of ice at 0°C to 1 gm of water at 0°C temperature. This much amount of heat is required to overcome the force of attraction between the particles of solids. Thus, when ice melts without changing the temperature, it absorbs latent heat and surrounding atmosphere becomes cooler. But when water becomes ice without changing the temperature, it gives out latent heat and surrounding atmosphere becomes hotter. When liquid water changes to vapour state i.e. steam without changing the temperature, it absorbs latent heat and surrounding atmosphere becomes cooler.

2. **Heat capacity and specific heat:** *The heat capacity* is defined "as the amount of heat (calories) needed to change the temperature of a given mass of the substance by 1 °C". The mass basis unit of heat capacity is cal/g °C while on volume basis it is cal/cm^3 °C.

   "*The specific heat* is defined "as the ratio of quantity of heat required to raise the temperature of one gram of the substance through 1 °C w.r.t that of heat required to raise the temperature of pure water by 1°C".

   The specific heat is expressed in terms of unit of cal g^{-1} °C^{-1}. The specific heat of water is 1.0 cal g^{-1} °C^{-1}whereas that of most minerals found in soils is nearly 0.2 cal g^{-1} °C^{-1}.

3. **Thermal conductivity:** Thermal conductivity is defined "as the quantity of heat transmitted through a unit length of a substance per unit cross-sectional area per unit temperature gradient per unit time". It can be expressed in terms of units of cal cm/sec-cm^2-°C or cal/sec-cm-°C. Compaction of the soil, content of water present, mineralogical composition of the soil, texture & structure, and organic matter content of the soil.

   Increase in water content increases the thermal conductivity of soil due to increase in the area through which heat flow. Increase in bulk density due to increase in soil compaction also increases the thermal conductivity due to increase in the contact area between the soil particles through which heat can be conducted.

   "In general, the thermal conductivity of soil varies in the following order: sand> loam > clay > organic soil".

   Fourier law of heat conduction (1822) - "It states that the quantity of heat (Q_h) flowing through the soil column under steady state condition is directly proportional to the cross sectional area of the soil column (A), difference in temperature between the two points, to the time and inversely proportional to the distance between the two points (x)".

$$Q_h \propto \frac{(T1 - T2)AT}{x}$$

$$Q_h = K_h \frac{(T1 - T2)AT}{x}$$

$$\frac{Q_h}{AT} = K_h \frac{(T1 - T2)AT}{x}$$

$$q_h = \frac{Q_h}{AT} = K_h \frac{(T1 - T2)AT}{x}$$

$$q_h = K_h \frac{\Delta T}{x}$$

Where q_h is the heat flux; $\frac{\Delta T}{x}$ = Temperature gradient; K_h = heat conductivity

4. **Thermal diffusivity:** It is defined "as the ratio of thermal conductivity to the heat capacity on volume basis". It is a measure of the rate of change in temperature with time.

   The thermal diffusivity can be expressed in terms of units of cm^2/sec.

5. **Albedo**: "The fraction of the incoming radiation that is reflected from the earth's surface is called as Albedo".

Albedo depends on many factors viz. soil colour, soil surface and the angle that rays of the sun makes with the Earth surface on being striking. More incidental rays are reflected back in morning as well as in evening than at other times of the day i.e. albedo is higher, and also during winter days than in summers as the sun is lower in the system. The soils absorbing higher amount of solar energy have lower albedo. Thus, the dark coloured soils will absorb more radiation and reflect less. This is the reason for lower albedo values of moist soils which are comparatively darker than drier soils. Also, the organic soils or the soils with higher organic matter will have lower values of albedo than the mineral soils. Similarly, rough surface have lower values of albedo than smooth surface.

6. **Thermal regime of soil:** It usually includes heat flux (movement) in soil, the thermal properties of soil (heat capacity, specific heat, thermal conductivity and thermal diffusivity) and heat exchange between soil and air.

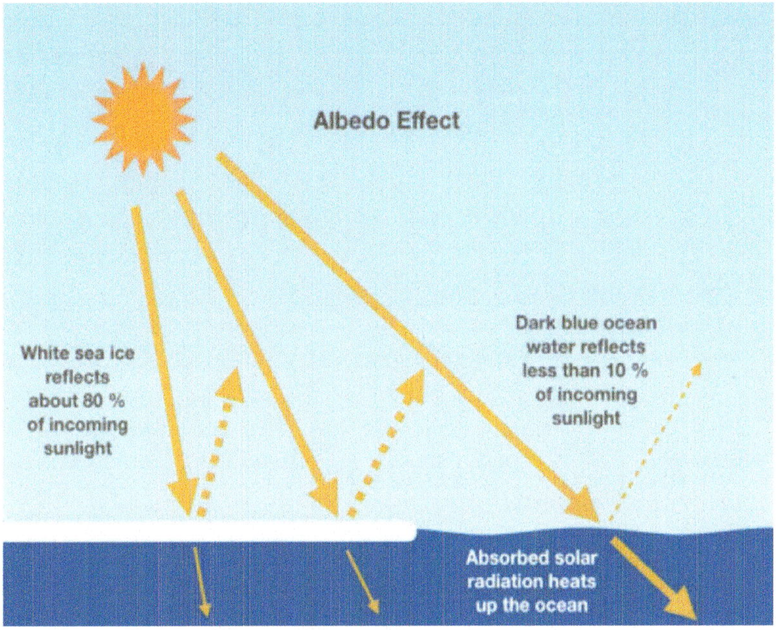

(Source: https://arctic-news.blogspot.com/p/albedo.html)

MODES OF TRANSMISSION OF HEAT

The heat may be transmitted through the soil by the process of

1. Conduction
2. Convection and
3. Radiation

Conduction: "It is the process in which heat is transmitted from hotter part of the system to colder when in contact with each other by internal molecular energy transference without any transference of material particles".

The movement of the heat is due to transfer of kinetic energy of the molecules of a body by numerous intermolecular collisions. Heat in solids is mainly transmitted by conduction.

Convection: "It is the process in which heat is transferred from one place to another place by actual movement of heated molecule from the place of higher temperature to lower". Liquid and gases are generally heated by the process of convection.

Radiations: "It is the process in which heat can be transmitted from hotter body to the colder body without intervention of material medium". The radiant heat energy travels in straight lines with the speed of light and is reflected or refracted,. By the process of radiations, heat reaches the surface of the earth from the sun.

FACTORS AFFECTING SOIL TEMPERATURE

The factors that affect soil temperature are categorized as:

1. **Environmental Factors:** The environmental factors that may affect the soil temperature are given below:

i) ***Solar Radiation*:** The amount of radiations reaching the surface of the soil may vary with the angle the sun makes with the earth, its latitude, season and time of the day, slope of the land and its angle, altitude from mean sea level, and interception of radiations due to air, water vapour, clouds, dust, smoke, fog (atmospheric condition), plants or mulch on soil.

ii) ***Latitude*:** Equatorial regions receive the maximum sun's heat as it makes more or less close to 90 degree angle in the sky than the sun's heat at polar latitudes i.e. north or south poles where the elevation of the sun is lower in the horizon.

iii) ***Season:*** The sun is at lower elevation during the winter season as compared to that in summer season. Hence, less radiation from the sun reaches the earth's surface during winter days than during summer days when the elevation of the sun is steeper. Thus, accounting for temperature fluctuations due to the change of the season in soil.

iv) ***Time of the day*:** At noon, higher amount of sun's energy reaches the earth's surface as the solar elevation is steeper (right angle) than that of the elevation in the evening and morning time when it is at lower angle in the horizon. Thus, accounting for the daily temperature fluctuation in the soils.

v) ***Steepness and direction of the slope*:** A sloping surface facing the sun at noon receives more heat energy as the sun rays fall at a right angle to it. In the Northern Hemisphere, the sun rays strike the steep south-facing slopes more near to the perpendicular than north-facing slopes.

vi) ***Atmospheric conditions*:** The air, water vapour, clouds, dust, smoke, fog etc. suspended in the atmosphere may insulate the solar radiations reaching the soil surface. Hence, during the cloudy day, or when the atmosphere is polluted by smoke or fumes or suspended with dust particles, more of the incoming solar radiation is reflected back to atmosphere. Thus, more radiation reaches the earth's surface in hot and dry regions than in humid ones.

vii) ***Evaporation*:** Heat is consumed when water is converted in to its gaseous state (endothermic process). To change 1.0 g of water to its gaseous state, 540 cal of heat is required. This heat energy consumed is lost from soil and returned to atmosphere, thus lowering the temperature of the soil.

viii) ***Condensation*:** Heat is released when water vapour condenses (exothermic process), therefore, soil temperature increases.

ix) ***Thawing (melting) and freezing*:** Thawing of ice is an endothermic process i.e. heat is absorbed (80 cal/g). Thus, the surrounding atmosphere becomes cooler with the loss of heat energy from the soil. Freezing on the other hand, is the exothermic process, will release this energy back to the soil.

x) ***Rainfall*:** Rainfall has a cooling effect on soil due to its low temperature,. Thus when the rain water percolates in to the soil, it rapidly cools the soil.

xi) ***Insulation*:** Insulation due to vegetative cover or its canopy, mulching, snow, clouds and fog has considerable influence on environmental temperature. Thus during summer months, the soils covered/insulated by vegetation are cooler than the non-insulated soils. This is due to the fact that most of the radiations from the sun are reflected back to the atmosphere because of interception due to vegetable resulting in less radiation to reach the earth's surface.

xii) ***Vegetation*:** Soils covered by vegetation have low temperature as more of the incoming radiations are reflected back to the atmosphere (higher albedo), decreasing the incoming solar radiation through interception by foliage, enhancing the latent heat of vaporization by transpiration and the use of energy for photosynthesis.

2. **Soil factors:** The soil factors that may affect its temperature are:

   i) ***Reflection of radiation through the surface of soil:*** There is a varying degree of reflection of radiations from the soil surface depending on the condition of the surface soil. For eg. if the surface of the soil is covered with fresh snow, then there may be around 70-90% of reflection of the incoming radiations, reflection from the sand dunes may vary from around 30-60%, from water surface it is 3-10 % while wet soils without vegetation have the reflection of 10 %.

   ii) ***Radiation from the soil:*** Out of total radiant energy absorbed by soil, a part of it is radiated back to the atmosphere and warms it. This radiation is called as *terrestrial* or outgoing radiation. It is a continuous process and thus a considerable amount of energy is radiated by earth to warm the atmosphere. It has been observed that during cloudy days that the radiant energy emitted by the soil do not goes to the atmosphere due to insulation by clouds and thus, the surrounding atmosphere around the soil remains warmer.

iii) ***Conduction and convection of heat from soil to the atmosphere:*** The movement of heat in the soil takes place from the warmer region to cooler one through the process of conduction and convection. A considerable amount of heat is lost from the soil, if atmospheric air above soil is relatively cooler.

iv) ***Heat fluxes in soil:*** The soil surface becomes warmer during day than the layers underlying it, therefore the heat flows downwards from surface to lower layers. However, the reverse occurs during the night, as the surface soil layer cools down due to decrease in temperature of the environment and interior of the soil remains warmer, with the resultant flow of heat from lower layers to top layer.

v) ***Specific heat of soil:*** The soil possessing more specific heat resists sudden temperature change. Therefore, wetter soils require more heat to increase its temperature. The soils with higher clay content have higher porosity especially the micro porosity that can hold more water and remain cool while light sandy soils which have low water holding capacity due to low porosity, warm up more quickly.

vi) ***Thermal conductivity of soil:*** The compacted soil conducts more heat due to more area of contact between the soil particles and conductivity of the soil is very high. The dry soil has less thermal conductivity due to decrease in the area of contact between the individual particles. With the soils becoming wetter, the area of contact between the particles increases and thus the thermal conductivity increases.

vii) ***Texture:*** Conductivity of heat is more in sandy soils than that of fine textured soils due to bigger size pores and more contact area of the particles of the former. The organic peat and muck soils have the least conductivity due to its extremely small size of particles. Thus, thermal conductivity increases with the increase in the size of mineral particles of the soil.

viii) ***Bulk density:*** The soils with high bulk density (Compacted soil) have higher capacity to conduct heat than the loose soil due to their more contact area. Tillage on the other hand loosens the soil and makes the surface soil poor in conduction of heat to lower layers due to less contact area and also due to discontinuity of the pores through which the heat flows.

ix) ***Structure:*** Soils possessing granular or crumb structure has less conductivity due to less area of contact of the particles because of more aggregation than the soils having platy or blocky structure.

x) ***Moisture content:*** Moist soils have high specific heat than the dry soils. Thus, the thermal conductivity of the moist soils is more due to more absorption of radiation.

xi) ***Soil colour:*** Colour of the soils highly influences the conductivity of radiations. Dark coloured soils have less albedo as most of the radiations are absorbed by it and very few is reflected back to the atmosphere. Thus, the thermal conductivity of dark soils is more.

xii) ***Biological activity:*** Degradation of material due to biological agents leads to the release of energy in the form of heat. Thus, the soils dominated with high microorganisms like that of organic soils, warm and moist soils etc, has high conductivity than that of the soil which are devoid of microorganisms

HEAT FLOW IN SOIL

In soils, heat flow takes place almost by the process of conduction. Although the heat flow can also takes place by convection and radiation but is of little importance in soils.

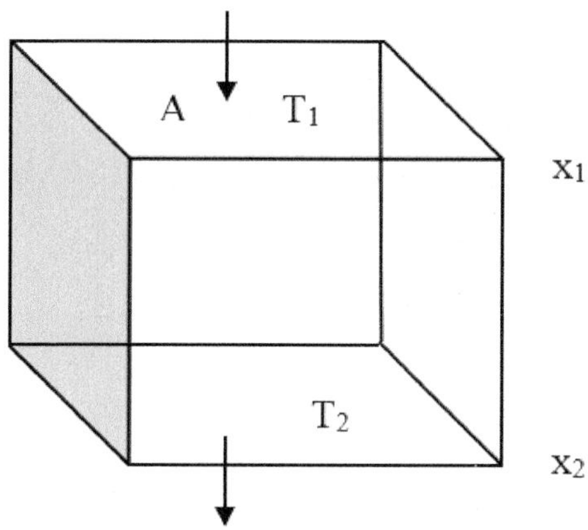

In *conduction,* heat flows by transfer of momentum of individual faster moving molecules at the hot end to their slower moving neighbouring molecules at the cooler end by collision. Heat flow by *convection* occurs due to differences in density of water as a result of temperature gradients in soil profile and vaporization of water molecules from the hotter end of a partially saturated soil pore and their condensation at the cooler end where it releases latent heat. The *radiative* transfer of heat in soil occurs in empty pores as thermal radiation.

Fourier's law described the heat flow in a soil through conduction. According to this law, "amount of heat transmitted per unit time (dQ/dt) in a unit area (A) of a block of soil (Fig. above) is directly proportional to the ratio of temperature difference (T_1-T_2 = dT) between two parts of the block to its thickness (x_1-x_2 = dx) i.e. dT/dx".

$$\frac{dQ}{dt} \propto A \frac{dT}{dx}$$

$$\frac{dQ}{dt} = -kA \frac{dT}{dx} \qquad (1)$$

$$\text{or } \frac{dQ}{Adt} = -k\frac{dT}{dx}$$

$$\text{or } J_q = -ki \qquad (2)$$

Where

J_q - flux density of heat,

dQ - amount of heat transferred in time dt across the area A,

k - thermal conductivity and the driving force,

dT/dx, = i, is the temperature gradient in the x-direction.

From equations (1 & 2), the thermal conductivity is:

$$k = \frac{J_q}{i} = -\frac{dQ}{(dt\ A\ i)} \qquad (3)$$

INFLUENCE OF SOIL TEMPERATURE ON PLANT GROWTH

There is a direct and indirect influence of soil temperature on the growth of the plant through change in soil moisture, aeration status of the soil, structure, biological activities, decomposition of plant residues and the availability of plant nutrients.

1. **Direct effects:** Soil temperature directly influences the germination of seeds, plant roots proliferation and growth of microbes. An optimum temperature of 10-35°C is required to germinate the seed, above and below which the seed germination is affected. The plant growth is initiated only when the minimum or optimum temperature is reached and the rate of growth increases up to the optimum temperature followed by a decline at the maximum tolerable temperature.

Crop roots become stunted and reduce branching when grown under extreme low or extreme high temperatures as soil's ability to withheld water is reduced and the roots cannot be able to withdraw water and thus the nutrients from the soil. It also depends upon the plant species and their tolerance to adapt to low and high temperature. The roots of the plants adapted to low temperature starts decaying if grown under high temperature conditions and thus increases the mortality rate of the plants or it may also be damaged by microorganism and are susceptible to diseases. The temperature congenial for the growth of roots of the most crop species is 20-25 °C and is often lower than that for shoot growth.

The micro-organisms activities is highest at a temperature between 25-35°C. The decomposition rate is faster which resulted in the production of humus that may increase the soil health which in-turn will benefit the plant in its growth. However, at low temperature, toxic substances are produced by the microorganisms that may be fatal for the plants.

2. **Indirect effects:** The soil temperature indirectly influences the soil moisture, structure, soil development, and their microbial activities and availability of nutrients.

a) ***Soil moisture:*** Temperature causes the soil water to move in the soils through its movement in the form of vapors'. The difference in vapor pressure or moisture content between any two points in the soil causes water (vapour phase) to move in the soil. Temperature is thus of significance in supplying moisture to the plant in case of deficit. For eg. in sandy soils, water may move upwards in vapour phase due to rise in temperature or it may percolates below root zone when in excess. In the night, the vapours condenses and thus become available to plants.

b) ***Soil structure:*** Increase in soil temperature causes desiccation of clay minerals leading to the development of deep wide cracks, thus eventually lead to breakdown of the soil structure. The soil structure may also be affected by alternate wetting and drying of the soil mass due to changes in soil temperature.

c) ***Microbial activity and availability of nutrients:*** Microbes thrive well at the optimum temperature of 25-35°C, below and above which their activities are highly affected. The reduced activities of the microbes may reduce the nutrient assimilation and its uptake from the soil. Also, microbes help in making the nutrient elements in to its oxide form, which is the preferred form of nutrients by the plants. Microbial activities also enhance the decomposition of organic matter and helps in stimulating the various processes such as mineralization, ammonification, amminization, nitrification, phosphate solubilizing etc. Hence, in-turn greatly influences the availability of nutrients.

d) ***Soil development:*** Temperature is an active factor which is important for most of the processes taking place in the soil. Hence, it determines the type of soil to be formed during weathering.

MANAGEMENT OF SOIL TEMPERATURE

Soil temperature can be managed by adopting the practices such as mulching, vegetation, irrigation etc.

1. **Mulching:** Mulch acts as a buffers for soil temperature. Crop mulch helps in lowering the temperature of the soils as it reflects back the incoming solar radiations. Poly-ethylene mulch, on the other hand, raises the temperature of the soil during the day as polyethylene is trans-

parent only to the incoming solar radiations. Mulching may also help in conserving the soil moisture by keeping the soil relatively cooler.

2. **Vegetation:** The vegetation intercepts with the solar incoming radiations and also entraps the radiation going out from soil and thus, helps in decreasing the temperature fluctuations within the soils. The temperature of the atmosphere above the crop is lower than on the soil surface.
3. **Tillage:** Tillage may directly affect the pore-size continuity and its distribution which alters the retention of soil moisture, and thus the soil temperature. The variation in soil temperature can be caused by building or planting on ridges and furrows. There is a large variation in soil temperature on ridges and furrows due to faster heat conduction, thus resulted in lowering the soil temperature. In soils prone to floods with poor drainage conditions, building ridge improves the soil aeration and thus manages the soil temperature. A tilled soil is generally warmer during the day time as tilled soil layer has higher porosity and low thermal conductivity due to discontinuity of the pores than the compacted or untilled soil.
4. **Soil compaction:** Thermal conductivity is high in a well aggregated soil due to its less area of contact. However, a poorly aggregated soils or a compacted soil has less conductivity of temperature due to its more contact area. Therefore, soil temperature changes are less observed in compacted sub-surface soils than the surface soils due to its high thermal conductivity.
5. **Irrigation and drainage:** Irrigation increases the soil water content that controls the solar radiation due to its absorption as the water has high specific heat and prevents the further heat loss to the atmosphere. The conduction of heat is faster in a wet soil than that of dry soil. Heat conduction is more in the surface layer due to its direct contact from the atmosphere than the sub-surface layer. Irrigation helps in maintaining the soil temperature. The soil temperature can be raised by draining the waterlogged soils and can be stabilized by irrigation.

Chapter 15

Soil Classification

The orderly arrangement of the objects (soils) with some logical sense is termed as soil classification. Soil classification is primarily based on differentiating characteristics of soils for their identification and study. For example, soils are classified as sandy, loamy or clayey soils based on their textures. For classifying the individuals of a large and widely varying population (e.g. soil), it is useful to group the individuals into classes, and further into higher classes. This kind of grouping is called a multi-categoric system of classification. Therefore, the individual soils are grouped into classes of lower category (soil series) which are further grouped into classes of higher categories (soil order).

Purpose of soil classification: The purpose of soil classification is:

i) To recognize properties of the soils
ii) To organize knowledge leading to economy of thoughts
iii) To know the relationships of different components of the soils and their principles for being classified
iv) To establish different categories and groups for practical and applied purposes so as to
    - Predict their behavior in different regions,
    - Knowing their potential use,
    - Estimating their production per unit area, and
    - Transfer the farm technology from research fields to that of farmer's field

EVOLUTION OF SOIL CLASSIFICATION SYSTEMS

Earlier system of classification of soils was based on economic, physical, chemical, geological and physiographic characteristics of soils. Since the early system of classification were based on a single character; therefore, these systems were of limited utility.

CLIMATIC CLASSIFICATION

This classification is given by "Vilenski" who recognized five temperature classes viz. polar, cold, temperate, sub-tropical & tropical and five moisture regimes classes viz. humid, semi-humid, medium, semi-arid and arid. But this classification failed to fit the soils all over the globe.

Recent system of soil classification

1. Dokuchaiev's Genetic system: In 19th century, a Russian Pedologist, V.V. Dokuchaiev - the founder of modern pedology- developed a soil classification system emphasizing on soil genesis. Based on Genetic system of classification, Dokuchaiev (1900) categorize the soils into - *Normal, Transitional* and *Abnormal* soils which were later named as *Zonal, Intrazonal* and *Azonal* soils, respectively.

Zonal soils: Those soils which have a fully developed mature soil with more or less in equilibrium with the climatic and environmental conditions viz. temperature, rainfall, vegetation etc. are termed as zonal soils. The soils having varying parent material but formed under similar climatic and biosphere has the comparable characteristics. Eg. are Sierozems, chestnut, podzols, laterites etc.

Intrazonal soils: These soils occur within a zonal area and the characteristics of soil are influenced by local conditions like parent material, topography, drainage status, soluble salts concentration, etc. For eg. Calcimorphic and Hydromorphic soils.

Azonal soils: The soils which possess poor profile or indistinct horizonation because of time as limited factor are termed as azonal. These soils include young/ immature soils without horizon differentiation. This may also be due to the parent material or adverse condition, e.g., slope. The soils formed by alluvial and colluvial material belongs to this group.

One of the fundamental law, formulated by Docuchiev and Sibirtezev is the law of adoptability of soil types of the globe to define natural conditions but there are many factors and geomorphological features that makes it difficult to classify all soils on the basis of climate alone.

2. ***Marbut's Morpho-Genetic System:*** Marbut (1935) accepted the concept of Dokuchaiev's Soil type, but renamed it as Great Soil Group. He realized that the Dokuchaiev's approach for soil classification had an inherent limitation which lays an undue emphasis on climate and vegetation rather than on intrinsic properties of the soils. Therefore, he introduced the concept of soil classification based on intrinsic properties of soils rather than on soil forming factors. His classification was based on the iron-alumina and lime contents and divided the Zonal Soils into two classes: *Pedalfers* and *Pedocals*. The *pedalfers* were presumed to occur in areas of high rainfall having excess leaching and accumul-ation of

iron and aluminium oxides. The *pedocals* were assumed to occur in areas of high evaporation having deficit of water and accumulation of calcium carbonate.

Limitations: The limitations of the genetic systems are:

i) The two highest categories (Order and Sub-Order) are described on the basis of genesis and not on the properties of the soils.

ii) The concepts and definitions of the highest category, i.e., the Order, in terms of soil properties are not clearly defined.

iii) The concepts and definitions of Great Soil Group are based on the qualitative parameters such as environmental factors rather than on the properties of the soil.

iv) Many of the soils properties are described under virgin soil conditions which are destroyed during cultivation and thus the classification of such arable soils becomes uncertain.

Soil Taxonomy: A Comprehensive System

A New Soil Taxonomy for classification of soil was developed in lieu of the limitation of the old classification system. The US Soils Survey Staff in cooperation with many other institutions have been working on this system since 1951 to develop/arrive at a classification system that narrows the differences in different view-points. The 7th approximation of a comprehensive system of soil classification (Soils Survey Staff, 1960) was published in 1960. The system was officially adopted by USA along with many other countries, including India, Iran, Belgium and Holland. A supplement of 7th approximation appeared in 1964 taxonomy.

ADVANTAGES OF SOIL CLASSIFICATION

The modern classification of soils i.e. soil taxonomy is based on morpho-genetic system which relies on those soil properties that exist today. The most important criteria to distinguish the soil is its properties besides being genesis. The system has advantages over the earlier systems in respect of the following.

1. Unlike the old Genetic classification system which was based on past properties of soil rather than the properties of the soil that was found recently, the soil taxonomy is based on measurable soil properties. Efforts have been made to define all classes in terms of soil properties that exist in a recent time.

2. It considers all such properties which are the outcome of soil genesis. Hence, soil genesis forms the backbone of the comprehensive system. But it does not appear in the definition of soil taxa.

3. The common definition of a class of taxonomic system is type.

4. The words used in nomenclature of taxonomic system were coined mainly from Greek and Latin languages which are most logical and help in relating the place of taxon in the system and in making interpretations.
5. A new category, i.e., subgroup has been added to define the central concepts of great groups and their intergrades in order to express and recognize more clearly that soils are in continuum and show gradual change in many properties.
6. In earlier system of classification of soils, the emphasis was given on the differentiating soil colour as the characteristic but soil taxonomy is absed on physical, chemical and biological properties for characterizing the soils for eg. Moisture, temperature, colour, texture, structure etc.

Structure of Soil Taxonomy: The system has six categories of classification. From the highest to the lowest level, the categories are grouped into two, i.e., higher and lower categories.

STRUCTURE OF SOIL TAXONOMY

ORDER (12)

(distinguished by presence or absence of diagnostic horizons which are greatly affected by climates)

SUB ORDER (63)

(differentiated by additional soil properties like soil moisture, temperature and textural features)

GRATE GROUP (300)

(Identified on the differentiating sub surface soil horizons and features such as Accumulated clays, Accumulated sesquioxides, Accumulated humus & presence or absence of diagnostic layers i.e. plinthite, fragipan, duripans that restricts water movement and root penetration)

SUBGROUPS (2400)

Intergrade Central (Typic) Extra grade

FAMILY (4500)

Soil properties that are most important for plant growth such as texture, mineralogy, pH, temperature, precipitation, permeability etc. are used to differentiate family

SERIES (several soil series)

No pedogenic significance, represents prominent geographic names of river, town or area where series was first recognized eg. Balsmand series

1. Higher categories

i) *Order:* The highest category in the system

ii) *Suborder:* Resembles like great soil groups of the Genetic system

iii) *Great Group:* A basic category based on diagnostic subsurface horizons

2. Lower categories

i) *Subgroup:* A new category that define the central concepts of great groups

ii) *Family:* A practical category for making predictions and land use plans

iii) *Series:* The lowest and most specific category

Order (12): This is the most generalized structure of taxonomy which is represented by presence or absence of major diagnostic horizons. All soils of the world fits into one of the 12 orders. Six of the orders are differentiated by the presence or absence of diagnostic horizons or other features. These features are characteristics of major processes of soil formation which are greatly affected by different climates.

Sub-order (63): Suborder are distinguished largely on the basis of additional soil properties which suggest genetic homogeneity and horizons resulting from differences in soil moisture, temperature, and dominating effects of chemical or textural features. Suborder is distinctive to each order and is not interchangeable to other orders.

Great group (240): Great groups are sub-division of sub orders. Diagnostic horizons are the primary foundation which depends largely on the basis of differentiating soil horizons and soil features. The differentiating soil horizons include those that have clay accumulations, presence of iron and /or humus compounds and presence of diagnostic layers viz. duripan, fragipan, plinthite that may restrict the water percolation and also the penetration of roots. The differentiating soil features are properties of clays, soil temperature, and major differences in content of calcium, magnesium, sodium, potassium, gypsum, and other salts.

Sub-group (> 1000): Each soil great group is divided into three kinds of subgroups, one representing the central (typic) segment of the soil group, a second which as properties that tend towards (intergrade to) other orders, suborders, or other great groups, and a third (extragrade), which has properties that prevent its classification as typic or an intergrade to another soil category. More than 1200 sub-groups are recognised.

Family: Family is based that soil properties which has the significance to the growth of plants or behaviour of soils for engineering purposes. The soil properties used include texture, mineralogy, reaction (pH), soil temperature, precipitation, permeability, thickness of horizon, structure, and consistency. About 4500 families have been identified so far.

Series: Each family contains several similar soil series. Soil series have narrower ranges or characteristics than the soil family. Soil series is of no pedogenic significance but represents a prominent geographic name of river, town or area near where the series was first recognized. Soil series are differentiated on the basis of measurable and mapable soil characteristics. The properties of the soils such as colour, texture, structure, consistence, thickness, reaction (pH), and number and arrangement of horizons in the soil pedon and their chemical and mineralogical properties are similar in the same series. Soils of a given series must have about the same kinds, thickness, appearance, and properties of horizons.

Diagnostic horizons of mineral soils: Diagnostic horizons are used to differentiate among soil order, suborder, great groups, and sub groups. A diagnostic horizon is defined as one, formed through pedogenic processes and having distinct features that can be described in terms of measurable soil properties. A number of diagnostic horizons have been defined in Soil Taxonomy. Based on their location in soil profile, diagnostic horizons are categorized into surface and subsurface.

The diagnostic surface horizons are called *epipedons*. The epipedons are simply the upper most soil horizons and include the upper part of the soil darkened by organic matter. Nine epipedons, viz. folistic, histic, melanic, mollic, anthropic, umbric, ochric, plaggen and grossarenic are recongnized. The epipedons viz. mollic, ochric and umbric are commonly observed in India.

The diagnostic subsurface horizons are called *endopedons*. The endopedons includes the lower part of the soil where soil materials accumulate. Nineteen endopedons viz. argillic, natric, agric, spodic, sombric, cambic, kandic, oxic, sulphuric, salic, placic, albic, glossic, calcic, gypsic, duripan, fragipan, petrocalcic, and petrogypsic are recognized. Out of these nineteen, eight endopedons viz. argillic, natric, cambic, kandic, oxic, salic, calcic, and gypsic are commonly observed in India.

The features of epipedons and endopedon are commonly observed in India are described as:

Epipedon (Surface horizon)		
Mollic	:	Thickness of mollic epipedon is more than 25 cm having dark coloured, friable soil horizon with >50% saturated with bases and having stable structure. It contains more than 1% organic matter with munsell colour values darker than 3.0 when moist and 5.0 when dry, and munsell chroma less than 3.5 when moist. For sandy soils, the mollic epipedon may be as shallow as 18 cm.
Ochric	:	A surface horizon that is thin and lighter in colour having Munsell colour value >3.5 when moist and >5.5 when dry. It contains low organic matter which is common in Aridisols, Entisols, and Inceptisols. It contains less than 1 % organic matter.
Umbric	:	A surface horizon like the mollic, but has base saturation less than 50%, however dominantly saturated with H^+. The horizon has high C:N ratio. The horizon remains wet for more than nine months in a year.
Histic	:	It contains around 20-30 % organic matter which depends largely on clay content of the soil. It remains water saturated for 30 days at some times of the year.
Anthropic	:	A horizon similar to mollic epipedon but contains over 250 ppm of citric acid soluble P_2O_5.
Plaggen	:	This is a man-made surface horizon that is more than 50 cm thick, created by years of manure (FYM) addition.
Endopedon (Sub-surface horizon)		
Argillic	:	It contains silicate clay enriched horizon formed by illuviation (accumulation at subsurface) of clay. The fine clay minerals are carried down by percolating water and is deposited on ped faces. In general, a B-horizon contains at least 1.2 times as much clay content as some horizon above, or 3 per cent more clay content if eluvial layer has < 15 per cent, or 8 per cent more clay if eluvial layer has > 40 per cent clay.
Agric	:	This layer is formed just under the plow layer and has accumulation of clay or humus content of atleast 15 by volume of the soil.

Epipedon (Surface horizon)		
Natric	:	This is similar to argillic horizon but is saturated with sodium ions to the extent of 15 % of the total cation exchange capacity. This horizon has prismatic or columnar structure caused by shrinking of soil mass due to reduction of water that sets horizontal tensions.
Spodic	:	This horizon has the accumulation of sesquioxides and organic matter with and without organic matter that gets illuviated to lower horizon with or without Iron. This horizon is generally formed in cold temperate and humid regions
Cambic	:	The subsurface horizon with some little amount of argillic or spodic horizon but not enough to qualify them as either. Formed under conditions of fluctuating water table and occurs widely under different environment.
Oxic	:	These soils are highly weathered and consists of largely of 1:1 type clay minerals and sesquioxides with low charge density.
Albic	:	An eluvial or bleached E horizon of podzols and planosols. It has munsell color value of ≥ 5 when dry or ≥ 4 when moist. The color values showed the removal of clay and/or free iron oxides from the horizon.
		Diagnostic layers
Fragipan	:	Fragipan has high bulk density, brittle when moist and very hard when dry but do not soften on wetting. It can be broken with the hands. This layer has low content of organic matter, base saturation and almost neutral pH.
Plinthite	:	Very low humus content almost negligible. Alternate wetting and drying leads to hardening of the soil to iron stone hard pans.
Duripan	:	This is a subsurface horizon having 50% dominated by SiO_2. Air-dry peds donot slake in water nor in HCl but disintegrates in hot concentrated KOH solution.

DIAGNOSTIC FEATURES

Soil moisture regimes

1. **Aquic regime:** This is a reducing regime as soil remains saturated with groundwater or capillary rise.
2. **Arridic or torric:** Soil moisture control section in most year is dry in all parts for more than half of the time and soil temperature at a depth of 50 cm is above 5 °C.
3. **Udic:** Soil moisture control section is not dry in any part for as long as 90 days. This is common in soils of humid climate
4. **Ustic:** Intermediate between aridic and udic regime. The moisture is limited but gets available when conditions get suitable for plant growth

5. **Xeric:** This regime is typified in Mediterranean climates where winters are moist and cool and summer are dry and warm.

Soil temperature regimes

1. **Pergelic:** This moisture regime has soils with mean annual temperature of < 0°C
2. **Cryic:** Soils having mean annual temperature of > 0°C but < 8°C.
3. **Frigid:** Soils having mean annual temperature of < 8°C and difference between mean winter and summer temperature of > 5°C.
4. **Mesic:** Soils having mean annual temperature of 8°C but < 15°C and difference between mean winter and summer temperature of > 5°C.
5. **Thermic:** Soils having mean annual temperature of 15°C but < 22°C and difference between mean winter and summer temperature of > 5°C.
6. **Hyper thermic:** Soils having mean annual temperature of 22°C

Taxonomic soil orders: In 1960, Comprehensive System of Soil Classification, known as 7th Approximation (Soil Survey Staff, 1960) was produced. Ultimately in 1975, the system was brought out as Soil Taxonomy (Soil Survey Staff, 1975); its second edition was produced in 1999. It is the modern and most accepted method of soil classification. It has more stress on morphology of soils (soil profile properties) themselves rather than on the environmental factors. Primarily, it is based on the study of the soil profile which is the ultimate product of and reflection of all soil forming factors and processes. In this system, higher categories are order, suborder, great group, subgroup; and lower categories are family, series and types.

In soil taxonomy, there are twelve soil orders on the basis of formative elements including the newly proposed Andisols and Gelisols.

Name of order (USDA taxonomy)	Name of order (FAO)	Formative element in name of order
Aridisols	Ermosols, Halosols, Xevosols	Id
Alfisols	Halosols, Planosols, Luvisols	Alf
Entisols	Lithosols, Fluvisols, Rhegosols, Arenosols, Gleysols	Ent
Histosols	Histisols	Ist
Inceptisols	Gleysols, Andosols, Cambisols	Ept
Mollisols	Gleysols, Halosols, Planosols	Oll
Oxisols	Ferraisols	Ox
Spodosols	Podsols	Od
Ultisols	Acrisols	Ult

Name of order (USDA taxonomy)	Name of order (FAO)	Formative element in name of order
Vertisols	Vertisols	Ert
Andisols	Andosols	-
Gelisols		

The higher categories give general understanding of soils over large areas indicating world-wide relationships. Whereas the lower categories give local differences and are important for assessing productive capabilities of soils for utilization purposes.

1. Entisols: (recently developed soil/ young soils) These soils are very recently developed mineral soils (Young soils) with no diagnostic horizon other than an ochric or anthropic epipedon. The main feature of Entisols is slight degree of soil profile formation. The reason behind slight profile development may be due to

i) Inert parent material or

ii) lack of time for development of significant profile (A-C horizons).

iii) Unfavourable climatic conditions or topographic conditions

(Source: www.earthreview.org)

Some weathering may be seen with the little organic matter at surface in these soils but in-situ or transported parent material is little affected during formation of soils. These soils shows the variation just as that of their parent material from which it is formed, which may range from deposition of sand dunes to a variety of rocks. Some soils may also occur on young geomorphic surfaces, such as areas affected by floodplains and on steep slopes of hills or mountains where the soil formed at top may be carried to base of the slope through gravity as soon as it is formed. They are formed under different climatic conditions. In rocky humid or sub humid mountain regions of Himachal Pradesh, medium textured Entisols are common. Sand dunes of Rajasthan, Haryana, central Punjab are extremely sandy Entisols are observed. The fluviatile Entisols are also observed in the recent flood plains. Some soils under Entisols have stony texture with low fertility or poor drainage conditions for cultivation. However, most areas in alluvial bottom lands are used for growing varieties of grains and vegetable crops.

2. **Vertisols**: (L. Verto-turning): The vertisols are characterized by black or dark coloured soils having very high content (>30%) of swelling type clays. In dry seasons, deep and wide cracks develop (> 1 cm wide) in such soils with high bulk density between the cracks. When it rains, water runs into the cracks and a significant amount of material from upper part of the profile may fall off into the cracks before the whole soil becomes wet and results in a partial inversion of soil. Such processes account for the term invert which is used to characterize this order in general way. They are generally unsuitable for construction purposes, roads and also for agricultural due to development of deep and wide cracks and also due to change in volume caused by change in moisture content. These soils are sticky and can be rolled into the desired shape i.e. plastic when wet and becomes hard when dry. These soils are mainly found in India (black soils), North-East Africa, Australia and Southern USA. In India, they are dominantly observed in Madhya Pradesh, Maharashtra, Gujarat, Andhra Pradesh, Tamil Nadu and Rajasthan.

(Source: www.earthreview.org)

3. Inceptisols: ***(L. Inceptum-beginning):*** Inceptisols are soils representing early stage in soil formation. The profile development is more advanced in this order than that of Entisols (least developed) but less than that of Alfisols. In sub-surface horizon, some clay may accumulate but it is does not qualify for an *argillic* horizon. Similarly, they may have dark coloured surface horizon which may be limiting in depth, organic matter or base saturation to be qualified for a mollic epipedon (Mollisols). The inceptisols are usually not dry and have one or more of the diagnostic horizons (cambic, umbic, or mollic with low base status). They have developed rather recently due to alteration of the parent material to develop soil structure but without much leaching and accumulation of material in the sub soil. These soils occur almost throughout the world.

In India, these are found all over except in hot and arid regions. Most of the Inceptisols are agriculturally productive and provide excellent natural grazing grounds, but some have limited productivity because of poor drainage and/or sodicity. Most of the Inceptisols (e.g. Ustepts) of Haryana and Punjab under assured irrigation conditions produce wheat yield as high as 4.5 tonnes/ha.

(Source: www.earthreview.org)

4. Mollisols (L. Mollis-soft): This order includes some of the world's most important agricultural soils. These are soils of grasslands (tall or short) vegetation under subhumid to humid environment. These soils are darker in colour, developed profile and rich in bases. The surface horizon is rich in organic matter, non-acid, poor in phosphate and characterized by a mollic epipedon. The surface horizons have sphere like structure with granular or crumby peds and remains soft and friable even when soils are dry. Most of the mollisols have developed under prairie (grassy) vegetation.

The subsurface horizons, rich in illuviated clay (argillic), calcareous (calcic) or gypsiferous materials (gypsic) may be present, but are not definition of the Order. Their characteristic features result from decomposition of organic material, especially fine roots of grassy vegetation, producing some stable dark compounds.

(Source: www.earthreview.org)

Mollisols are predominantly observed in central USA, and from east to west across the Russia and in the central South America. In India, such soils are dominantly observed in the Tarai region of Uttar Pradesh and Uttaranchal. These soils are also found in some regions of H.P., northern Bihar, and in Maharashtra and M.P. The mollisols are inherently the best agricultural soils of the world. In India, these soils produce optimum yields both under irrigated and unirrigated conditions with minimum inputs. Under arid climatic regions. these soils are put to use for grazing purpose only. In humid regions, crops like wheat, maize, sorghum, and vegetables, etc can be grown.

5. ***Aridisols (L. Aridus-dry):*** These mineral soils are mostly found in dry climates with high ground water table, and the soil layers are dry through out the most of the year except in regions where water irrigation water is available. Leaching of the soil constituents is very less due to the limited water availability. Surface epipedon is *ochric* very low organic matter, light in colour due to arid climatic condition.

These soils have higher amount of calcium carbonate (calcic), gypsum (gypsic), or soluble salts concentration (salic). The Aridisols widely occur in the arid and semi-arid climatic environment, especially in South Asia, north and north east Africa, Australia, south-west part of south America, northern America, South Africa and through east to west of Russia.

In India, these soils occur in some parts of Rajasthan, Gujarat, Haryana, etc. These soils do not support crops without irrigation, However, if irrigated for cultivation,they have risk of salinization. In non-gypsiferous landscapes of Haryana and Punjab, they are used for growing usually pearl millet n a year by following water conservation practices during summer monsoon period and growing the crop with stored moisture. The Aridisols have limitations of xerophytic climatic conditions and brackish groundwater which comes to surface due to high evaporative demand rendering the soils saline and unsuitable for plant growth, except for salt tolerant vegetation.

(*Source: www.earthreview.org*)

6. Spodosols (Gr. Spodos-wood ash): The soils which have a spodic subsurface horizon (white in colour), with an accumulation of amorphous organic matter along with oxides of iron and aluminium. These develop under cool, humid climates and coarse textured siliceous parent material which favours free leaching conditions. These soils can also occur under hot to warm and humid tropical regions where the parent material is sandy rich in quartz and ground water fluctuates. Illuviation of organic material along with oxides of Fe and Al is the dominating process underlying an *albic* (light in colour) horizon justifies reference to wood ash. These soils are formed in high rainfall and temperate areas. These soils are largely distributed in north-west Europe, northern and north-east USA, Canada, etc. In earlier system of soil classification, these soils were classified as Podzol, Brown Podzolic and Hydromorphic soils by recognizing the bleached ash coloured alluvial E-Horizon. Acid producing vegetation like pine, hemlock etc. favours the formation of these soils. The parent material is mostly dominated with silica. These are not very fertile soils due to lack of organic material and bases but can be make productive under proper fertilization.

7. Alfisols (Alfus-weathered): Alfisols are base rich, mineral soils of sub-humid and humid regions. The soils are characterized by a light coloured *ochric* epipedon over a clay-enriched *argillic* subsurface horizon (illuviated clay more in B horizon than A) having more than 35% base saturation. Alfisols are intensely weathered soils than the Inceptisols but comparatively less than the Ultisols. Profiles are also more developed than Inceptisols.

These soils are widely distributed in the USA, South America, Northern Europe, south East Asia, etc. In India, Red soils (Alfisols) are observed in different agro-ecological regions covering parts of states like Andhra Pradesh, Assam, Bihar, Arunchal Pradesh, Himachal Pradesh, Orissa, Karnataka, etc.

In the past, these soils were classified as Grey Brown Podzols, Planosols and Half Bog soils. The Alfisols are naturally fertile and Productive because of high base saturation, favorable texture and location in moist semi-arid to humid region. In India, they are used for growing crops viz. wheat, maize, sorghum, rice, mango, cashew, etc. The sodium rich Alfisols need amendments such application of gypsum.

(Source: www.earthreview.org)

8. Ultisols (L. <u>Ultimum</u> - latest) : These are usually moist soils and developed under warm, humid and sub-tropical climates. The Ultisols are similar to that with Alfisols the only difference is that the base saturation less than 35% which is due to more intense weathering. They also have the dominance of silicate clay with *argillic* or *kandic* endopedon. Except wetter soils, the sub surface horizons are commonly red or yellow in colour due to accumulation of free oxides of Fe.

These soils are moistly found in south-east America, South America, south east Asia, central Africa and coastal India. These soils are found in Kerala, Tamil Nadu, Orissa, Assam and other north eastern states of India. Due to low fertility and base saturation, these soils are limited to agricultural use.

These soils are mainly used for forestry, but may be used for agriculture when the soils are fertilized and amended with adequate lime. In tropical regions, Ultisols are used for pineapple, sugarcane, coffee, coconut, rubber, etc. These soils can be cultivated for vegetables with adequate fertilization

(Source: www.earthreview.org)

9. Oxisols (Fr. Oxi-oxides): These soils are the extremely weathered mineral soils of humid tropics in the classification system. The presence of deep sub-surface horizon (oxic endopedon) dominated with kaolinite clay minerals and oxides of iron and aluminum is the characterizing and distinguishing feature of these soils. Extremely strong weathering and high leaching leds to the removal of silica from the silicate minerals in this horizon leaving behind more amounts of oxides of Fe and Al. Most Oxisols have brick red colour but some are also yellow or grey in colours. These soils are dominantly observed in central Africa, north-east South America and south east Asia. In tropical India, these soils are found in Kerala, Tamil Nadu, Orissa and Karnataka. In the past, these soils were classified as Laterites, Latosols and Ground water Laterites.

(*Source: www.earthreview.org*)

The Oxisols are chemical degraded soils and need careful management for agricultural use. The soils are mainly used for grazing and forestry and also for growing crops like coffee, rubber, cocoa, sugarcane and tropical fruits like pineapple, coconut, jackfruit, etc. Shifting agriculture is common practice in the areas of Oxisols. This practice is acceptable when the shifting cycle is more than 20 years, but due to increasing population and demand for more food, the cycle has been reduced to 3 to 5 years which results in serious problems of soil erosion by water and nutrient depletion.

10. Histosols (Gr. Histos - tissue) The histosols are also known as Bogs, moors, peat or mucks. The Histosols are soils which are formed on organic soil materials. These soils are organically rich with thick and peaty horizons (*histic* epipedon-tissue like surface horizon) that develop in lower topographic positions in permanent water saturated environment. It may lead to formation of process of accumulation of peat which happens when production of organic matter by native vegetation is more than its decomposition in the soil. The amount of organic material required for a soil to be qualified as Histosol depends on the clay content. The organic matter content of these soils is at least 20% organic matter if the clay content is low (or nil) and 30% organic matter if the clay content is > 50%. The Histosols occur in some parts of northern parts of Europe, northern USA, Canada, and North west Asia. In India, the Histosols are found in some locations of Kerala, Tripura,and Andaman and Nicobar Islands.

(Source: www.earthreview.org)

In the past, these soils were calssified as Peat, Muck, and Bog soils. Based on the degree of decomposition of humus, the Histosols have been subdivided into four orders: Hemists, Fibrists, Saprists, and Folists. Most of the Histosols, located in depressions and plain areas have natural vegetation. The Folists are well drained soils. The Hemists have moderately decomposed organic matter and are used as woodland/rangelands.

***11. Andisols (from Japanese, dark surfaced)*:** Andisols were first recognized in Japan and named as Volcanic ash material. The Andisols are found to form from the deposition of volcanic material in recent geological times. These soils are not highly weathered mineral soils. The upper horizons of these soils are dark in colour. Previously these soils were classified as Inceptisols.

(*Source: www.earthreview.org*)

The typical characteristics of these soils are the high content of allophone which gives them a very low bulk density and fluffiness, especially in B-horizon. These characteristics of Andisols make them highly fertile and easy to cultivate. Since the Andisols are dominantly developed on volcanic ash, or materials derived from it, they are mostly observed around the Ring of Fix in the Pacific, including Japan, Indonesia, New Zealand and Philippines. They are also found in the 'Rift Valley' of Africa, Kenya, Western coast of USA and South America. In India these soils may be found in islands of Andaman and Nicobar. On steep slopes, the Andosols may have limitation of mechanized farming. However, Andisols in favorable climatic conditions are intensively cultivated and are thickly populated. Andisols are generally stable and resist water erosion because of high infiltration and permeability rates. When dry, these soils are susceptible to wind erosion. Major limitation in these soils is their high P fixing capacity which makes P in an unavailable form, therefore, crops showing little response to the application of phosphorus can be grown on Andisols.

12. Gelisols ***(Latin gelare, to freeze):*** The Gelisols is the most recently introduced soil order of soils occurring in the extreme northern hemisphere. The Gelisols are the soils with gelic materials underlain by permafrost. Diagnostic horizons may or may not be present in Gelisols as thawing and freezing play an important role in their evolution. Permafrost layer affect their pedogenesis as it forms a barrier for the percolation of the soil solution downwards. Gelisols are commonly observed in areas of permafrost which include northern USA, Canada, Greenland, Russia, and other areas in the extreme northern hemisphere. They may be observed in higher Himalayas with permafrost conditions (ice caps in extreme north of J & K and Sikkim). The Gelisols are not cultivated because of climatic conditions. They may have natural vegetation.

(source: www.earthreview.org)

Chapter 16

Soil of India

MAJOR SOIL GROUPS OF INDIA

1. Black soils of India
2. Laterites and lateritic soils
3. Red soils
4. Alluvial soils

(Source: www.mapsofindia.com)

Other Important groups of soils in India

1. Desert soils
2. Forest/Mountain soils
3. Tarai foot hill soils
4. Peaty and marshy soils
5. Salt affected soils

1. Black soils of India

Genesis:

- Formed from Deccan Basalt trap rocks either in-situ or transported parent material
- Black soils may also be formed on parent material Granite or Gneiss containing lime and feldspar
- Topography consequently drainage is an important soil forming factors

Gilgai microrelief

(Source:https://en.wikipedia.org/wiki/Gilgai)

(Source:https://www.fickr.com/photos /soilscience/5097001159)

Occurrence:

- Black soils occur in area with high rainfall climate consisting of semi-arid to sub-humid type.
- Climate is hot and dry summers followed by rainy period with rainfall of 40-100 cm followed by mild to moderate winters.
- Major area of black soils are Maharashtra, M.P. and parts of A.P., Gujarat and Tamil Nadu.

Morphology:

- Black soils are clayey soils (35-60 %) with smectite type minerals which impart swelling and shrinkage properties.
- Gilgai microrelief (naturally occurring topographical feature that results in a unique undulating ground surface. These undulations develop from deep seated swelling pressures within the underlying expansive soil layers) and **slicken sides** (lens shaped blocks formed within the soil slide past one another during this swelling action. This created the polished surface called slicken side) are seen in these soils.
- The structure of black soils are crumby or granular. The black soils have impeded drainage and are of low permeability.

Classification:

These soils are classified under Entisols and Inceptisols and the deep black soils in Vertisols.

2. Red soils (Tera Rosa, Italian word means red earth)

Genesis:

The parent materials for the formation of these soils are granite, gneiss and other metamorphic rocks either on the place of their origin (in-situ) or may also be transported parent material to other places. These soils are formed under properly drained conditions and the fundamental processes such as Eluviation and Illuviation are mainly dominant for the formation of these soils.

Occurrence:

Most occur in states of A.P., Assam, Goa, Bihar, Daman and Diu and parts of Kerala, Maharashtra, Karnataka and major parts of T.N. and West Bengal.

MORPHOLOGY OR CHARACTERISTICS

- Major features of red soils are light textured, friable structure, slightly acidic to slightly alkaline, absence of lime concretions and free $CaCO_3$.
- Dominant mineral is ***Kaolinite***

- Low water and nutrient holding capacity and thus deficient in N, P & K
- The colour may vary from Reddish to yellow due to coating of Ferric oxides on soil ped surface (Yellow colour due to hydrated ferric oxide)
- Fine-grained red soils are usually fertile while coarse grained red soils are less fertile.

Classification:

Most of these soils are classified under ***Alfisols***

(Source: www.bbc.com)

3. Laterites and Lateritic soils- (Latin word later means brick)

(Source: www.bbc.com)

Laterite is formed from the Latin word which mean brick. Thus, the name indicates the formation of brick like hard material probably at subsoil horizon. The term laterite was given by Bucknan (1807) with high dominance of oxides of Fe (ferruginous), vesicular material and unstratified/massive deposits observed in the hills of Malabar, Kerela, South India. The laterite is used for those Red soils that qualifies for oxic and/or Kandic subgroups and have low base saturation.

Genesis:

- Lateritic soils are formed due to eluviation of silica and enrichment with Iron and Aluminum oxides i.e. Laterization is the dominant soil forming process. In high rainfall areas, silica gets released from the surface and translocate to lower layers leaving behind Fe and Al at surface.

- At higher altitude, the Laterites soils formed may be transported down to lower levels or the base of the hill/mountains by water, air or gravity and cemented in to compact mass and are termed as low levels Laterites.

Occurrence:

- Well-developed Laterites are observed on the summits of hills of Deccan, Karnataka, Kerela, M.P. and Ghat regions of Orissa, Maharastra, Kerela and also in West Bengal, Tamil Nadu and Assam.

Morphological Characteristics:

- Characterized by low Silica:Sesquioxide ratio and also devoid of alkali and alkaline oxides.
- At greater altitude, these soils are usually pale, gritty (coarser in texture), shallow in depth and low in soil nutrients whereas those developed at low altitude, are fine-textured, rich in nutrients and organic matter.
- The pH of the laterites are generally low with low CEC.
- Kaolinite clay minerals are dominant. Gibbsite is commonly observed in Laterites.
- The soils are deficient in Phosphorous due to high Phosphate fixing capacity as Fe and Al-phosphates.

Classification:

- Classified in Entisols, Inceptisols and Alfisols

4. Alluvial Soils

The alluvial soil is given to those soil groups that have developed due to transport of material carried by water i.e. alluvium parent material of the climatic condition. They represents the most fertile soils for agricultural production.

(Source: www.bbc.com)

Genesis:

Alluvium soils are formed on transported parent material due to action of water, ice, gravity , wind etc. Geologically alluvium is divided into

- Khadar - Recent/new alluvium

- Bhangar – Older alluvium

(Source: www.bbc.com)

Occurrence:

- The alluvial soils are extensively distributed in the Indo-Gangetic Plains and Brahmaputra Valley and also occurs in the coastal and deltaic regions.
- Alluvial soils are distributed mainly in northern, north-western and north eastern parts of India including Punjab, Haryana, UP, Bihar, West Bengal and parts of Assam and Orissa.
- Some coastal alluvium soils are found along the sea coast of Gujarat, Maharashtra, Kerala, T.N., A.P., Orissa, W.B. and Goa

OTHER IMPORTANT GROUPS OF SOILS IN INDIA

1. Desert Soils

(Source: www.bbc.com)

- The texture of desert soils varies from sandy to loamy fine sand having lighter in colour ranging from light brown to yellow brown. The salts present are soluble with alkaline pH of more than 8.0. The nitrogen concentrations of these soils are very low due to high temperature and limited moisture.
- If the water is present in these soils, it may become fertile due to presence of phosphate and nitrates. These soils occur extensively in the state of southern Haryana, some parts of Punjab and Rajasthan.

- The soils of desert regions falls under order ' Aridisols and Entisols.
- The soil temperature remains very high almost all part of the year with an average temperature of 34°C.
- The weathering is very less due to less rainfall and high temperature. The particles of desert soils are larger in size due to less and slow weathering. The fine particles are taken away by winds to other places. Due to larger size particles, they may posseses higher number of macro pores due to which water drain off quickly, thus are mostly dry almost throughout the year.
- Desert soils are lighter in colour due to low amount of organic matter as the high temperature may oxidize the organic matter in to CO_2 and H_2O. As the climatic condition is not favourable for the growth of vegetation, so the soils have very less growth of plants with very less diversity of microorganism exists in the soils. These soils are often fertile but the due to lack of water, they cannot become productive.
- The presence or accumulation of alkaline earth carbonates is the striking feature of most desert soils with low clay content having pH ranging from 8.0 to 8.8.

New alluvium is sandy and light coloured whereas older alluvium is more clayey, dark coloured and contains lime concretions

Classification:

Genetic sytem: Azonal (Regosols)

Taxonomy: Aridisols and Entisols

2. Forest/ mountain soils

These soils were put under Zonal soil Order as per earlier soil classification, under Pedalfers that are soils with the following characteristics

- Soils of hill and mountain forest-grassland transition
- Light-coloured soils having podsolization as the dominating soil forming factor of timberland regions with coarse gritty texture
- Lateritic soils formed under warm and cool temperature condition such as those prevailing in tropical forests.

Characterictics

(Source: www.bbc.com)

- Soils are acidic in nature due to high rainfall and leaching of basic salts (Ca, Mg, Na and K) leaving behind the Fe and Al which imparts acidity to the soil.
- At higher altitude, the particles of these soils are coarse-grained and loamy and silty on low land soils as that in valley.
- In the snowbound areas of the Himalayas, these soils undergo high erosion through water wind ice etc. and are acidic with low humus content. However, low land valley soils are comparatively more fertile.

3. Tarai soils

(Source: www.bbc.com)

Soils having wet regimes and high water table condition for most part of the year. Tarai are the foot hill soils and extends in lines having varying widths at the base of Himalayas in J&K, U.P., Bihar and West Bengal.

Soils were derived from material carried downwards due to erosion from the tops of mountains. The Taria soils are classified under alluvial soils. Presence of more moisture leads to the luxuriant vegetation which may include tall grasses. Enrichment with organic matter is of special significance in horizon differentiation. Soils having mollic epipedon have been put in Mollisols.

4. Peaty and Marshy soils

Peat soils are found in water-logged humid regions having large amount of partially decomposed plant material/organic material including sphagnum moss and other acid-loving plants which has been formed around 10000 years in poorly drained wetland habitats.

With the advancement of decomposition of organic matter by microorganisms, it becomes fibrous to semi-fibrous with amorphous material. This may results into impeded drainage by blocking of pores of the soils and that to even on sloping ground. The oxygen gets depleted due to land completely covered with water, the organic matter decomposition retards and thus the peat begins to accumulate. These soils are rich in organic matter but deficient in phosphorous and potash.

(Source: www.hutton.ac.uk)

The peat which gets accumulate on the soil may cover the surface thoroughly just like a mulch or a blanket across undulating land and hills and is thus called blanket peat or blanket bog. The peat soils which accumulate in valleys and basins with high water accumulation, it is called as basin peat or raised bog.

SOIL FORMING FACTORS

PARENT MATERIAL	Dead and decayed debris of the plants
VEGETATION/ ORGANISMS	"Vegetation is dominated by Common heather, Hare's tail cottongrass, Peat moss (Sphagnum), Deergrass, Cross-leaved heath, Purple moorgrass and Common cottongrass. Decomposition is mainly carried out by fungi in the surface horizons. Due to the acidity, earthworms are absent and their role is taken over by the much smaller Enchytraeid worms. Bacteria that form methane are found in the waterlogged zones".
CLIMATE	Cool and humid environmental condition favours the formation of peat and marshy soils
TOPOGRAPHY	Peat soils are found anywhere from sea level to mountain plateaus but are generally limited to slopes < 15°.
TIME	For the formation of peat bog soils, it took several thousand years

5. Salt affected soils

(Source: www.shutterstock.com)

Saline soils are having high soluble salts of Ca and Mg with some Na and K. The soils become salt affected due to in-situ parent material, high temperature and arid environment. The salts travel from downwards to the surface due to evaporation and settles on the surface of the soil. These soils also developed in area of high groundwater condition. The electrical conductivity of the soils is >4 dS/m and ESP<15.

In Saline–sodic soils, there is a dominance of soluble salts as well as exchangeable Na^+, thus the electrical conductivity of these soils are more than 4 dS/m, ESP>15 but the pH of these soils is less than 8.5 due to high electrolyte concentration. High salt concentration makes the soil floccules. Therefore, these soils are treated with gypsum ($CaSO_4.2H_2O$) firstly and then leaching is followed otherwise Na^+ will disperse the soils and the problem of blocking the pores will arise.

To settle the salts to the deeper layers, surface and sub-surface drainage should be improved and water of good quality should be used for leaching the salts from the soil.

Another salt affected soil is the sodic soils, which has an EC <4.0dS/m, pH > 8.5 and ESP>15. The soils are treated with gypsum to reclaim the soils followed by 2 to 3 leaching with good quality water.

Chapter 17

Soil Reaction and Buffering Capacity

Soil reaction or pH is the very chief electrochemical property that may have an impact on physico-chemical and biological properties of the soil. Soil reaction refers to the acidity and basicity of the soil. Soil reaction is measured in terms of pH of the soil which -

- Indicates the chemical environment of soil.
- Affects soil properties- chemical, biological and indirectly even physical properties.
- Influences nutrient availability and activity of soil micro-organisms.
- Influences mobility of pollutants in soil.

EXPRESSION OF SOIL REACTION (pH)

Soil reaction i.e. pH is the relationship between H^+ (acidic) and OH^- (alkaline) ions in the soil suspension which indicates whether the soils are acidic or alkaline or neutral. The pH which is the measure of soil reaction is defined "as the negative logarithm of the hydrogen ion activity or the amount of active hydrogen ions present in the soil solution" and can be expressed in moles per litre.

$$pH = -\log (H^+)$$

Where (H^+) indicates the activity of hydrogen ions in moles per litre.

Danish chemist ***Sorensen*** (1909) suggested expressing H^+ acitivity conveniently by logarithmic scale. Acidity or alkalinity is expressed by pH ranging from 0 to 14.

The pH can be measured on the scale of 0-14. Neutral soils have pH of 7.0 (such as of pure water). With decrease in the pH value (from 7-0) acidity increases, and increase in the pH value from 7 to 14 will decrease the acidity.

(Source: https://www.alamy.com/stock-photo/ph-scale.html?sortBy=relevant)

Soil pH may be determined either calorimetrically or electrometrically. The colorimetric method of pH determination has a variation of 0.2-0.5 pH unit but is a simple, inexpensive and moderately accurate. In a calorimetric method, the soil is mixed with the dye and stands for an hour or so. The colour developed of the mixture is compared with the colour on colour chart.

In electrometric method, the electric pH meter is used which measures the concentration of active hydrogen ions through a glass electrode when come in to the contact with the soil solution. The method is most accurate but requires the thoroughly calibration of the meter with the standards. The initial cost of the equipment is relatively expensive, and requires one hour or more of contact of soil and water before the measurements are made. The accuracy of the method is about 0.1-0.2 pH units.

Importance of soil pH

1. The soil pH strongly influences the dissolution of various minerals. In strongly acidic soils, there is more solubility of minerals even to the level of toxicity especially those of aluminum, iron and manganese. The crop and its species grow well under strongly acidic condition includes tea, cranberries, blueberries and several timber plantations etc. However, crop like alfa alfa, barley, wheat, sugar beet can be grown in slightly acidic to slightly alkaline conditions due to their high requirement for calcium.

2. Soil pH has a considerable influence on the growth of plant due to the effect on the activity of microorganisms. Bacteria thrive well at a pH value of 6.5 to 7.5 i.e. in neutral range above and below which their activities are highly affected. Activity of bacteria is affected strongly in highly acidic condition whereas fungi can grow well at acidic condition. Similarly, high alkalinity with pH more than 8.5 also is fatal for the plant growth. Plants may even die at pH value of more than 9.0.

3. Majority of nutrient ions like iron, Zn, Cu Mn and phosphates get precipitated in the soil solution by $CaCO_3$ at higher pH.

Types of soil reaction: On the basis of concentration of hydrogen and hydroxyl (pH), the reactions occurring in soil are of three types i.e. acidic, alkaline and neutral.

1. **Acidic:** The soil reaction is generally acidic (pH < 6.5) in areas dominated with high amount of rainfall and warm temperature regime. The high rainfall increases the leaching of appreciable amounts of exchangeable basic cations viz. Na^+, K^+, Ca^{2+} and Mg^{2+} from surface soil to lower layers and the exchange complex is saturated by H^+ ions.
2. **Alkaline:** The regions dominated by low rainfall generally favours alkalinity of the (pH > 7.5) due to comparatively high degree of base saturation occurring in arid and semi-arid regions. The strong base such as Na_2CO_3 hydrolyses to give rise to alkalinity.

$$Na_2CO_3 \longrightarrow 2Na^+ + CO_3^{-2}$$

$$2Na^+ + CO_3^{-2} + 2HOH \longrightarrow 2Na^+ + 2OH^- + H_2CO_3$$

There will be greater dissociation of NaOH under alkaline condition than the carbonic acid leading to dominance of OH ions with the resultant increase the alkalinity of soil and soil pH rises to 9 or 10.

3. **Neutral:** The neutral soil reactions occur in regions where soil solution have equal amount of hydrogen and hydroxyl ions. Soils having pH in the range of 6.5-7.5 are considered as normal soil

SOURCES OF HYDROGEN AND HYDROXYL IONS IN SOILS

1. **Strongly acidic soils:** In strongly acidic soils (pH <5.0), most of the aluminum and iron get solubilized in soil solution. It may also be present in adsorbed form i.e. adsorbed by organic manure or is present in the form of aluminum hydroxyl cations. The contribution to increase the soil solution acidity is very less because of the strongly adsorbed H+ and Fe and Al ions by organic matter and clay content due to strong hydrogen bonding. When aluminum ions in the soil solution are hydrolyzed, it may forms the aluminium hydroxide ions and release of H^+ in the soil solution-

$$Al^{3+} + H_2O \longrightarrow Al(OH)^{2+} + H^+$$

The H^+ ion thus released during the reaction lowers the pH of the soil and contributes as a source of hydrogen in strongly acid soils.

2. **Moderately acid soils**: In moderately acid soils (pH 5.0 to 6.5), Al and H compounds also accounts for soil solution H^+ ions. The base saturation of moderately acid soils is slightly higher than strongly acid soils.

$$Al^{3+} + OH^- \longrightarrow Al(OH)^{2+}$$
$$Al(OH)^{2+} + OH^- \longrightarrow Al(OH)_2^+$$
(Aluminum hydroxy ions)

In soil solution, few aluminium hydroxy ions produce hydrogen ions by the hydrolysis reactions.

$$Al(OH)^{2+} + H_2O \longrightarrow Al(OH)_2^+ + H^+$$
$$Al(OH)_2^+ + H_2O \longrightarrow Al(OH)_3 + H^+$$

3. **Neutral to alkaline soils (pH ≥ 7):** Neutral to alkaline soils are neither dominated by H^+ nor by Al^{3+}. The exchange sites of the soils are saturated with Ca^{2+} and/or Mg^{2+} or other base forming cations. The di hydroxy aluminum ions are converted to insoluble gibbsite by the reaction follows:

$$Al(OH)_2^+ + H_2O \longrightarrow Al(OH)_3$$
(*Gibbsite***)** **(***Insoluble***)**

Factors affecting soil reaction:

1. *Nature of the clay colloids*: Clays possessing high cation exchange capacity (CEC) have tendency to adsorb and release more H^+ like montmorillonite as compared to clays having low CEC like Kaolinite. Thus, pH of the soil with montmorillonite is less than kaolinite.
2. *Concentration of soil solution:* With soils becoming drier and drier, soil pH decreases i.e. soil solution gets concentrated and number of H^+ per unit volume increases. Since soil reaction varies with the salt concentration, it is preferred to measure pH in soil suspension in 0.01M $CaCl_2$ or 1N KCl solution to avoid the fluctuation in pH measurements.
3. *Presence of* Al^{+3}*:* Presence of more Al^{+3} ions make the soil acidic and vice-versa.
4. *Percentage base saturation:* When the per cent base saturation of the soil is < 80, the soil becomes acidic and if it lies between 80-90 the soil is considered neutral. However, at base saturation of > 90 (low rainfall areas of arid soils), the soil reaction becomes alkaline.
5. *Decomposition of organic matter:* Micro-organism decomposes the organic matter resulting in formation of organic acids, CO_2, water and carbonic acid. Carbonic acid on reaction with Calcium and Magnesium to form their carbonates and soluble bicarbonates. The leaching of soluble bicarbonate makes the soil more acidic.
6. *Nitrogen fertilization:* Nitrogeneous fertilizers on hydrolysis releases H^+ in the soil solution which increases acidity. Nitrogeneous fertilization increases the rate of acidity development.

7. *Flooding:* The flooding has the different effect on increasing or decreasing the pH of the soil. Flooding the acid soils will increase its pH as it increases the solubility of Fe and Al oxides and releases hydroxyl ions in the soil solution while in alkaline soils due to decomposition of organic matter, there will the release of H^+ in the soil solution, with the resultant decrease in the pH of the soils. On flooding, most of the soils reach to pH of 6.5 to 7.5 within a month, irrespective of their original pH values.
8. *Soil management*: Cultural operation generally tends to increase the soil acidity. Consistent cultivation resulted in loss of basic cations from the soil by the process of leaching and more crop removal. Also, injudicious and indiscriminate use of chemical fertilizers continuously is responsible for changing the pH of the soil.

Availability of nutrients at varying soil reaction (pH)

There is strong relationship between soil pH, and the availability of plant nutrient (figure), as well as the activities of soil-organisms.

Strongly acidic soils (pH <5.5): In strongly acid soils, most of the macronutrients viz. Calcium, magnesium, potassium, phosphorous, nitrogen and sulphur along with micronutrient molybdenum are reduced. While most of the micronutrient's availability viz. Fe, Mn, Zn, Cu and Co is increased by lowering soil pH, even to the extent of toxicity to plants and micro-organisms.

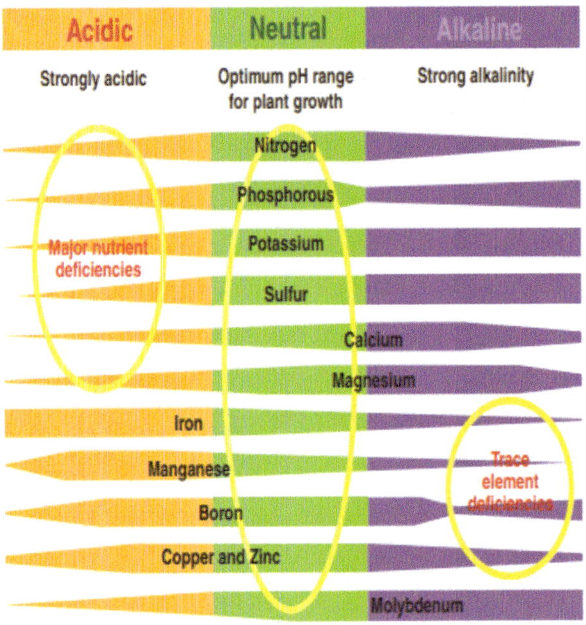

Soil pH and Nutrient Availability
(Source: Bluedale - http://www.bluedale.com.au)

Slightly to moderately alkaline soils (pH 6.0 to 8.0): Availability of Mo, N, K, Ca, Mg, Na and K is increased but Fe, Mn, Cu and Co availability is reduced

Strong alkaline soils (pH > 8.0):

Unavailability of nutrients like N, P, Fe, Mn, B etc. occurs in strongly acidic soils. Also, the trace element deficiency is usually observed.

The availability of different nutrients with pH is described as:

1. ***Nitrogen:*** The availability of N (nitrate) depends on the activity of nitrifying bacteria. These bacteria are most active at pH 6.5 - 7.5. Their activities are adversely affected if the pH falls below 5.5 and rises above 9.0.

$$\text{Organic nitrogen} \xrightarrow[\text{(Mineralization)}]{\textit{Bacteria, fungi}} NH_3 + H^+ \longrightarrow NH_4^+$$

$$\underset{}{NH_4^+ + 3O_2} \xrightarrow{\textit{Nitrosomonas}} \underset{\textbf{(Nitrite)}}{NO_2^-} + 2H_20 + 4H^+ \xrightarrow{\textit{Nitrobacter}} \underset{\textbf{(Nitrate)}}{NO_3^-}$$

2. ***Phosphorus:*** The availability of phosphorus is highest at soil pH 6.5 - 7.5. In strongly acidic soil, Fe and Al are present in abundant soluble amount which combine with phosphorus to form phosphates making phosphorus unavailable. The unavailability of phosphorus is called *phosphorus fixation*. The fixation of phosphorus also takes place when the soil pH is high. At high pH, phosphate combines with calcium and magnesium to form insoluble calcium or magnesium phosphate. The $H_2PO_4^-$ dominates in highly acidic soil (pH 4.0 -5.0). The concentration of HPO_4^{-2} starts increasing with increasing pH. At pH 9 and above, the PO_4^{-3} are available to plants. The $H_2PO_4^-$ and HPO_4^{2-} ions are considered to be more available than PO_4^{3-} ions.

Acidic pH:

$$\underset{\textit{(Soluble)}}{Al^{3+} + H_2PO_4^- + 2H_2O} \longrightarrow 2H^+ + \underset{\textit{(Insoluble)}}{Al(OH)_2H_2PO_4} \quad \textit{(Phosphorous fixation)}$$

Alkaline pH:|

$$\underset{\textit{(Soluble)}}{Ca(H_2PO_4)_2} + 2Ca^{2+} \longrightarrow \underset{\textit{(Insoluble)}}{Ca_3(PO_4)_2} + 4H^+ \quad \textit{(Phosphorous fixation)}$$

3. ***Potassium:*** The potassium of the soil does not have impact on the pH of the soils..

4. ***Calcium and magnesium:*** Calcium and magnesium concentration in in acid soils is very low. In alkaline soils, the Ca and Mg availability is high at pH value of < 8.5. However, their availability decrease at pH higher than 8.5.

5. ***Sulphur:*** The soil pH has no effect on the availability of sulphur. However, its availability is more soluble in acidic condition and lost as leaching.

6. ***Iron and manganese:*** The Fe and Mn are readily available in acidic soils. At low pH, the solubility of these micro nutrients increases. At pH below 5.5, the solubility of these nutrients becomes high which is toxic to plant growth. At pH range 5.5 to 7.0, Fe and Mn are present in soluble ferrous (Fe^{+2}) and manganous (Mn^{+2}) forms. In alkaline conditions, Fe and Mn are generally present in ferric (Fe^{+3}) and manganic (Mn^{+3}) states. Thus, in soils having pH 7.5 and above, iron and manganese become unavailable.

7. ***Boron, copper, molybdenum and zinc:*** The availability of boron, copper and zinc increases in the acid soils while their availability decreases in alkaline soils. However, availability of molybdenum decreases in acid soils while increase in alkaline soil.

POOLS OF SOIL ACIDITY

The acidity of the soils are classified in to three categories such as:

a) Active acidity

b) Exchangeable acidity

c) Residual acidity

a) **Active acidity**: It represents the activity of H^+ ion in the soil solution at any particular time. The active pools as well as the exchangeable pools of H^+ ions are in equilibrium with each other. Thus, this acidity which is most readily soluble in soil solution is the most important pool of soil acidity that may affects plant growth directly. Active acidity estimated by using electrode based pH meter.

b) **Exchangeable acidity**: The exchange acidity is linked with the Al^{3+} and H^+ ions adsorbed to the exchange sites of soils. Treating the soils with unbuffered KCl salt will released these ions into soil solution and K^+ may come on the exchange sites.

$$\begin{matrix} Al^{3+} \\ H^{+} \end{matrix}\boxed{} + 4KCl \rightleftharpoons K^{+}\boxed{} + AlCl_3 + HCl$$

(Exchangeable) *(Soil solution)* *(Exchangeable)* *(Soil solution)*

In moderately acid soils, the quantity of easily exchangeable aluminum and hydrogen is quite limited. However, the lime requirement neutralizing the exchangeable acidity is still much higher than that required for neutralizing the same amount of active acidity. Exchangeable acidity is maximum for Smectites group i.e. montmorillonite and vermiculite and minimum for Kaolinite clay minerals.

c) **Residual acidity**: The acidity left after the active and exchangeable acidity gets neutralized is referred to as residual acidity. Those H^+ and Al^{3+} ions which gets bound by organic matter and clay minerals in non-exchangeable form are the residual acidity.

Management of soil acidity/soil alkalinity

Increasing and decreasing the pH value can be done by several methods:

- **Flooding**: Flooding will increase the pH of the acid soils to near neutral range while decrease the pH of the alkaline soils to neutral range. The reduction of Fe and Al will release the hydroxyl ions in the acid soils that may increase the pH while the oxidation of organic matter present in the alkaline soils will lead to production of hydrogen ions which decrease the pH of the soil. This change in soil reaction upon flooding the soils exists up to the point when soils are anaerobic i.e. flooded and as soon as aerobic conditions prevails, again the soil attains its original soil pH.
- Proper selection of crop is also necessary for the optimum growth of the plant. Crops most suited to be grown under anaerobic conditions are preferred under flooding condition. Crops like Rice, Taro etc. can be grown under anaerobic conditions.
- **Organic matter**: The organic matter addition in the soil helps to manage problems associated with high soil acidity as well as alkalinity. The cation exchange capacity of the soil is highly influenced by organic matter, thus addition of organic matter will lead to increase the exchange sites on the soils which increases the base saturation. With the increase in base sites on the soil particles, the amount of acidic cations decreases. In alkaline soils, decomposition of organic matter releases organic acids and H^+ ions which may decrease the pH of the soil to near neutrality. In addition, organic matter may form the chelates, with aluminium ions which may reduce the availability of aluminium ions and thus the soil acidity.
- **Additions of wood ash**: Wood ash also increases the bases in the soils just as that of organic matter and thus form chelates with aluminum.

- **Addition of amendment**: Amendments commonly used for acidic soils are liming materials such as limestone (calcium carbonate), dolomite (calcium/magnesium carbonate), hydrated lime (calcium hydroxide), and quicklime (calcium oxide). Calcium and magnesium silicates are also used as liming agents. Addition of gypsum may decrease the pH of the alkali soils.

Buffering of the soils

Soils can resist the changes in pH when large amount of acidic or basic materials are added. So, the capability of the soil to resist the changes in pH of the soil is called as buffering capacity of the soil. Hydrogen is present in the soil in the solution form as well adsorbed on the exchange sites of the clay minerals of the soils. The pH measures the concentration of these hydrogen ions in the soil solution. Soils differ in the number of surface sites are able to accommodate hydrogen ions. The buffering capacity increases as the exchange sites on the soil increases.

The soils having large clay content or organic matter/humus content possesses high exchange sites that can be able to hold hydrogen ions and are able to resist the change in pH. When lime is added to the soil to neutralize soil acidity, an equal proportion of H^+ ions are released from the soil exchange sites to solution form and the sudden change in the pH of the soil.

Clayey soils are highly buffered than loams, which in turn are better buffered than sands.

Importance of soil buffers

- Changes in soil reactions have a direct influence on availability of nutrients in the soils as some nutrients are deficient at given pH while others are available at toxic level at a given pH. So buffering capacity of the soils help in preventing the sudden change in the pH which are harmful to the plants and microorganisms.
- Knowledge of soil buffering capacity helps to estimate the required amounts of amendments such as lime or sulphur to bring about the desired change in soil pH.

Factors affecting buffering capacity of the soil

- **Texture**: Clay loam soils have higher capacity to buffer the change in pH of the soil as compared to sandy loam or silt loam due to higher cation exchange capacity of the former. Thus, the soils dominated by higher amount of small size particles (<0.002 mm) have higher capacity to resist the change in pH of the soil solution than the larger size particles.

- **Type and amount of clay minerals**: The soils dominated by montmorillonite or vermiculite have higher capacity to buffer than the soils with kaolinite or illite clay minerals due to the more exchange sites of the former. Black soils have higher capacity to resist the changes in pH of the soil than res or lateritic soils.
- **Amount of organic matter present in the soil**: Higher the organic matter, higher will be the capacity of the soil to resist the changes in pH. Organic matter has much more negative charges than clay content, so has more cation exchange capacity. Thus organic soils, have higher buffering capacity than the minerals soils.

Chapter 18

Soil Colloids

The physico-chemical properties of the soils are largely governed by the soil colloidal particles. According to classification system given by varying agencies, clay particles of the soil fraction are those particles that are less than 0.002 mm in size. Out of these clay particles, those particles which are less than 0.001 mm in size are called as soil colloids i.e. soils colloids are those whose particles are very small in size i.e. smaller than 0.001 mm in diameter. Therefore, all colloids are clay particles but all colloids are not clay. Colloids are classified as

1. **Inorganic colloids** (clay minerals, hydrous oxides): These colloids exist in the soil in large proportion.
2. **Organic colloids**: It includes largely the dead plant and animals generally the organic matter designated humus. Organic colloids are considered highly reactive chemically under favourable soil and water conditions and have more impact on soil physicochemical properties than as compared to inorganic colloids.

   Humus is amorphous organic substance while clay minerals are mostly crystalline structure (with the exception of few as amorphous). The crystalline structured clay minerals possess definite physical properties and must have specific chemical composition i.e. possess definite characteristic form.

SOIL COLLOIDS AND ITS SPECIFIC PROPERTIES

1. **Surface area**: As the soil colloidal particles exhibits very small size, therefore, possess large specific surface (i.e. surface area per unit mass or volume). Colloidal clays possess large external and internal surface area. The external surface area is considered to be atleast one thousand times more for colloidal clays as compared to coarse grain particles per unit mass. The alumino-silicate clays for eg kaolinite have surface area of 10 m^2/g which is entirely due to external surfaces, however, montmorillonite clay mineral have surface area of 800 m^2/g which arises due to large external as well as internal surfaces.

2. **Surface charge**: Clay colloids carry either negative charge or positive or both but in most of the soils, colloids possess net negative charge. However in highly acidic soils, positive charge dominates due to protonation effect. The presence of charge density on soil colloids helps in adsorption and desorption of cations/anions on the soil clay particles and thus regulating physic-chemical properties of the soil.

The clay minerals are negatively charged although positive charge also exists on the clay but the net charge is –ve and that may arise due to the following factors:

i) **Isomorphous substitution**: The net negative charge on clay colloids arise due to exchange of cations of higher valency with that of lower valency having more or less of similar size in tetrahedral (SiO_4^{4-}) and octahedral sheets of clay minerals. During the crystallization process, Si^{4+} ions are exchanged by Al^{3+}, and Al^{3+} by Fe^{3+}, Fe^{2+}, Mg^{2+}, or Zn^{2+} in tetrahedral and octahedral sheets of clays. The substitution by lower valent cation with higher valent cation resulted in excess of the negative charge on the site of the soil.

For eg

$$Si_2O_4 \xrightleftharpoons{\text{Isomorphous substitution of Si by Al}} SiAlO_4^-$$

The substitution of higher valent cations for lower valent cation yields net positive charge on clay colloids. For eg.

$$Mg_3O_2(OH)_2 \xrightleftharpoons{\text{Isomorphous substitution of Mg by Al}} Mg_2AlO_2(OH)_2^+$$

ii) **pH dependent charges** arises due to ionization of hydrogen ions or by protonation. The H of the hydroxyl group of the -Al-OH portion of the clay dissociates in alkaline soils and thus leaving behind an extra negative charge on the oxygen group ($-Al-O^-$). However, in acid soils, the protonation of $-Al(OH)^+$ resulted in the excess positive charge. Thus, the dissociation and protonation of H^+ depends upon the acidity and alkalinity of the soil, and vary with variation in the soil reaction i.e. soil pH.

3. **Adsorption of cations**: The cations such as H^+, Al^{3+}, Ca^{2+}, and Mg^{2+}, etc., are attracted to the clay colloids surfaces as they possess negative charges which ultimately give rise to elctrostatistical diffuse double layer.

4. **Adsorption of water**: Water being dipolar in nature, thus a large number of water molecules are also associated with the soil colloidal particles. Water molecules are attracted to the adsorbed cations

and cations get hydrated. Some water molecules are held in the internal surfaces of the colloidal particles and results in the swelling of colloidal particles.

5. **Cohesion**: It is the process in which similar type of the colloidal particles are attracted towards each other through force of attraction lying between them. For example: attraction of one water molecules to another and sticking together of one clay particle to another and so on.
6. **Adhesion**: Adhesion is the process of joining together of the colloid particles of differing types. For example: attraction of colloidal particle with foreign objects like plough, tillage implements etc.
7. **Swelling and shrinkage**: Some clay has the capability to swell when water is made available such as montmorillonite/ vermiculite and shrink when dry. After a prolonged dry spell, soils high in smectites e.g., Vertisols, are often characterized by wide and deep cracks. During rain or when irrigation is applied, initially these soils due to wide cracks allow the water to infiltrate rapidly, but later on due to swelling of clay minerals, these soils become more impervious than the soils dominated by koalinite, chlorite, or fine grained mica (Illite). Vermiculite is intermediate in its swelling and shrinking characteristics.
8. **Flocculation and Dispersion:** Flocculation process involves the coagulation of the individual particles of clay colloids to form floccular aggregates. The nature of adsorbing calcium ions decides the degree and permanence of flocculation. For example, higher valent cation and exceptionally hydrogen tend to increase flocculation. Dispersion on the other hand involves the separation of the individual particles from one another which may be carried out by potassium and sodium ions in the soil-water suspension. Thus, depending upon the cations present in a soil, it may be either in a flocculated (aggregated) or in a dispersed (massive) state. Sodium saturated clays have a thick electric double layer surrounding the ion due to their large hydrated radius forcing the ions to apart from each other, that means the clays remain in suspension. While on the other hand, divalent, trivalent and tetravalent eg. Calcium, aluminum, silicon suppresses the double layer and cause flocculation..

Types of soil colloids

Soil colloids are classified into two types

1. Inorganic soil colloids
2. Organic soil colloids

1. **Inorganic soil colloids**: The inorganic soil colloids are the layered aluminum and silicate associated clay minerals; Fe and Al oxides and their hydroxide, allophane and other associated amorphous silicates, and clay sized primary minerals
2. **Organic colloids**: Humus is an organic colloid.
   i) **Layered Alumino-silicate clays:** These alumino-silicate clays are categorized as *phyllosilicates* clay minerals since they possess leaf-like or plate like structure. These clays are made up of silicon dominated tetrahedral sheet and aluminium and/or magnesium octahedral sheet. The silica tetrahedral sheets consists of silicon atom which is surrounded by four atoms of oxygen and thus the arrangement is called as silica tetrahedron.

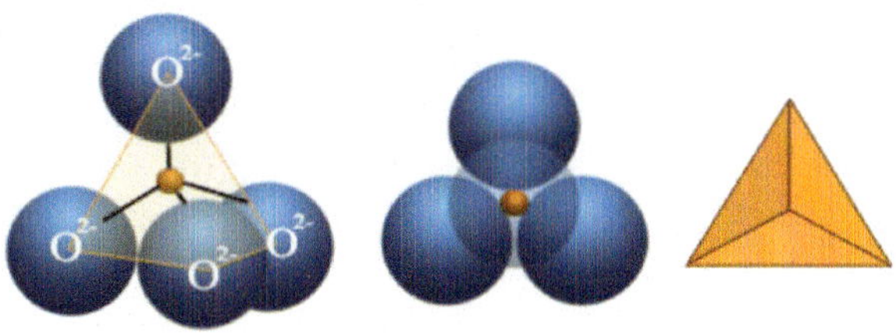

Silicon tetrahedron

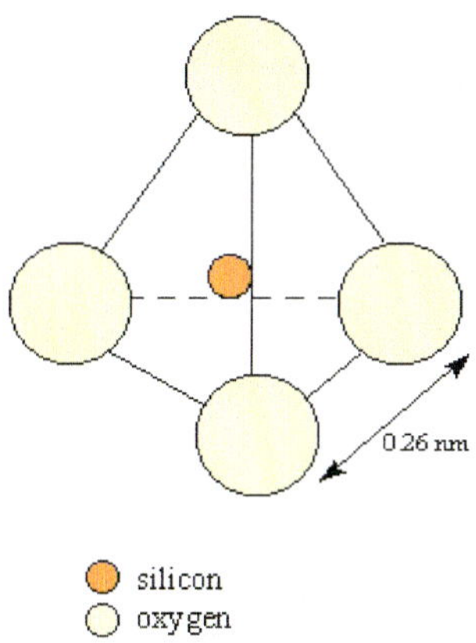

Structure of tetrahedral units

(Source: www.expertmind.com)

The octahedral unit made up of aluminium and/or magnesium ions is bonded by six O-atoms or OH group and called aluminium (or magnesium) octahedron having eight sided configuration (Fig). These octahedral units are horizontally joined together to each other to form the sheets of octahedral units. The octahedral sheet dominated with aluminum ion is designated as *di-octahedral* sheet, whereas the one dominated by the Mg ion is designated as *tri-octahedral* sheet. These sheets are called as di-octahedral and tri-octahedral as the two aluminium ions of the di-octahedral sheet satisfy the same negative charge from surrounding oxygen and hydroxyl ions as the three magnesium ions of the tri-octahedral sheet.

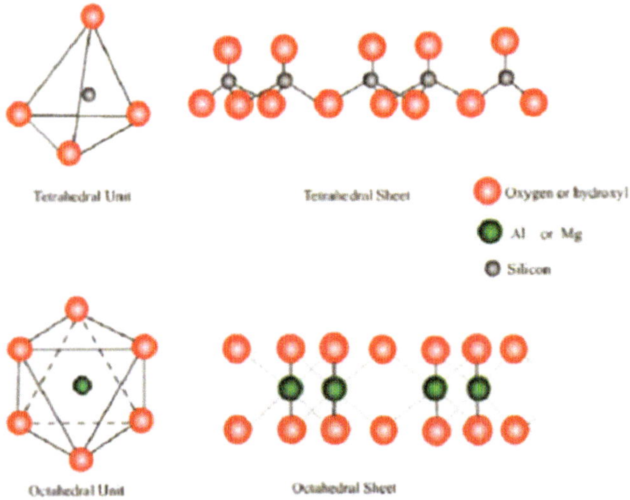

(*Source: www.expertmind.com*)

The structure of silicate clay minerals are composed of sheets of tetrahedral and octahedral units which are linked together with the crystals units with the oxygen atoms into different layers by strong hydrogen bonds due to which structure is fixed and no expansion takes place between layers when clays are wetted.

TYPES OF CLAY MINERALS

Silicate clay minerals/clays are designated into 3 different types based on the type of tetrahedral and octahedral units linked together to form sheets:

i) Clay minerals with 1:1 type arrangement
ii) Clay minerals with 2:1 type arrangement
iii) Clay minerals with 2:1:1 or 2:2 type arrangement

1. **Clay minerals with 1:1 type arrangement:** One silica tetrahedral sheet when combine with one aluminum octahedral sheet forms, it is called as 1:1 type (Fig). For examples: Kaolinite, Hallosite, Nacrite, and Dickite.

The kaolinite is one of the most significant example of 1:1 type di-octahedral clay mineral. These tetrahedral and octahedral sheet are held together tightly by oxygen atoms in a layer of a kaolinite crystal, which are respectively shared by the silicon and aluminium ions in the tetrahedral and octahedral sheets. The layers of the 1:1 type are joined together by strong H- bond, i.e. structure is more intense and water molecules cannot enter in to the interlayer space of the lattice and thus no expansion occurs between the layers. The water along with cation also cannot be able to enter the interlayer spaces of the lattice of these caly minerals, and thus the cation exchange capacity of these clay minerals is less. Also, the isomorphous substitution of higher valent cation with lower valent cation in Kaolinite mineral (1:1 type) is negligible and the negative charge in this arises mainly due to large external surface of the mineral. It is because of this reason, these clay minerals possess low cation exchange capacity.

The shape of Kaolinite crystals are hexagonal and size ranging from 0.10 to 5.0 µm. These clays show the properties of low plasticity, cohesion, swelling and shrinkage. The layer of the structure is strongly bonded to each other thus do not allow the expansion of the kaolinite nor do break down into extremely thin plates. Kaolinite is generally found in soils of Ultisols, Oxisols and to some extent Alfisols.

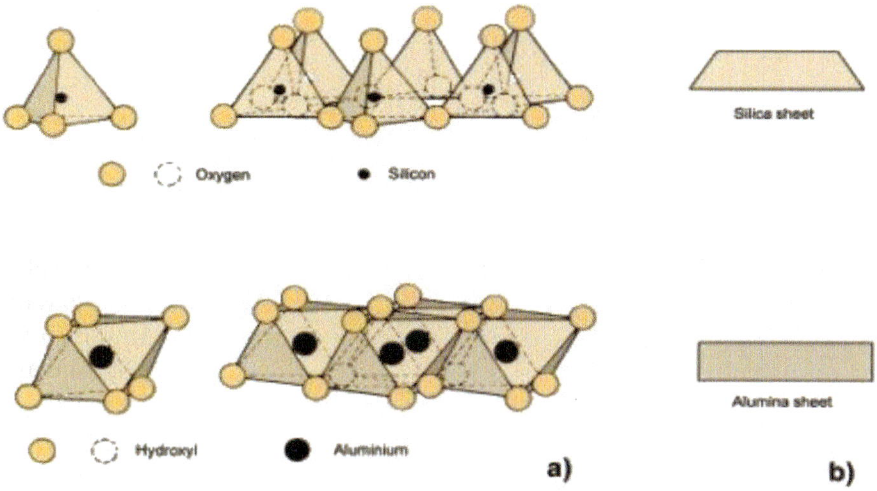

(Source: www.expertmind.com)

2. **Clay minerals with 2:1 type arrangement:** The 2:1 type of clay minerals constitutes octahedral sheet crammed between two tetrahedral sheets (Fig). They are classified in to two groups:
   i) **2:1 Type Expanding clay mineral:** Examples are Smectite group viz. Montmorillonite, Beidellite, Nontronite and Saponite minerals and Vermiculite mineral showing very less expansion)
   ii) **2:1 type Non-expanding clay mineral:** Example is fine grained micas (weathers to form Illite)

There is interlayer expansion in expanding type smectite group. As the water enters the inter-layer spaces of these minerals, they swells and apart the layer from each other. The most distinct example of expanding type clay minerals found in the soil is Montmorillonite. Other examples of expanding type clay minerals are Beidellite, Nontronite and Saponite.

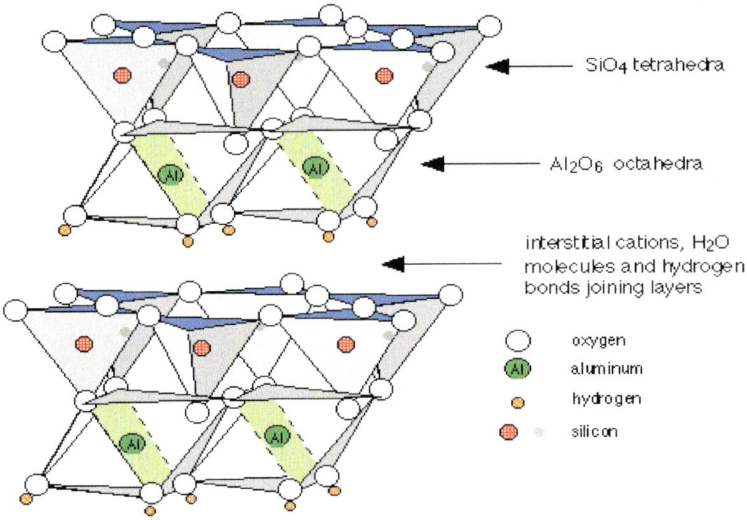

(Source: *www.expertmind.com***)**

Montmorillonite possess plate like structure which are loosely held to each other as there is less attraction between oxygen atoms of the lower and upper sheets of tetrahedral units This loose bonding creates a variable space between the layers of this mineral and thus is occupied by water and exchangeable cations. There is a large variation in the internal and external surface area of these minerals with the former exceeding much than the later. The isomorphous substitution of higher valent aluminum by lower valent magnesium in octahedral sheet and higher valent silicon by lower valent aluminium in tetrahedral sheet give rise to the high negative charge in these minerals. The cation exchange capacity of the montmorillonte clay mineral and swelling and shrinkage properties is very high. Thus under dry conditions, these minerals (Smectite dominated soils i.e. Vertisols) shows deep wide cracks and it becomes difficult to plough.

In 2:1 type non-expanding mica clay mineral, Al ions occupied $1/4^{th}$ of tetrahedral sheets due to which an excess of negative charge per formula unit in 2:1 layer aroused. The negative charge is being occupied by K^+ ions that occupy inter layer sites between the two adjacent 2:1 layers. Addition of water does not apart the interlayer lattice of these clay minerals, thus, the K ions between the unit layers cannot be able to exchange for another cations as it get snuggly fit in the interlayer lattice of the clay crystal, they are fixed. The K^+ ions at the external surface are only responsible for exchange for another cation.

3. **Clay mineral with 2:2 or 2:1:1 type arrangement:** In this 2:1:1 clay mineral, an Mg hydroxide tri-octahedral sheet is attached to the 2:1 type smectite mineral. It may contain Mg, Al, Fe hydroxide di-or tri-octahedral unit sheet. *Example*: Chlorite. This is also a non-swelling clay minerals as water do not enter in to the layer structure of the lattice.

i) **Hydrous Oxides of Fe & Al:** These clays are commonly dominated in highly weathered soil of tropics and semi-tropics and in significant quantities in soil of some temperate regions e.g. Gibbsite (Al_20_3. $3H_2O$) Goethite (FeOOH).

Some have definite Crystalline structure but other are amorphous. These clays not sticky & plastic as layered silicate clays.

ii) **Allophanes & Other amorphous minerals:** In many soils, these are significant quantities of colloidal matter that is ether amorphous or has a crystalline structure not sufficiently ordered to be detected by X-rays. As they lack 3-dimensional structurer, these-fore they are referred to as short range ordered minerals.

4. **Organic colloids (Humus):** The final product of the decomposition of organic matter is Humus which is an important constituent of soils. It forms the largest fraction of soil organic matter and plays dominant role in improving the soil productivity. Humus is complex substance formed by decomposition of plant and animals residues by soil micro-organisms. It is fairly resistant to further decay (in contrast to peat moss, which is relatively undecayed organic matter but which also has a high CEC), and thus forms stable organic component of the soil (but not as stable as clay). It is polymeric substance, dark coloured (black and dark brown to yellow), and nearly insoluble in water but soluble in NaOH or KOH solution. Because of their wide range of molecular sizes and properties, the humic substances are grouped into fractions:

   i) **Fulvic acid**: This is the most highly soluble fraction of the humus which is soluble in water, acids as well as alkali. Colour ranging from yellow- red to brown-yellow.

   ii) **Humic acid**: Humic acid is soluble in alkali only and insoluble in water and acids. It is dark brown to black coloured.

   iii) **Humin**: This humus fraction is the most highly insoluble fraction which is not at all soluble in acids and alkalis. This fraction of humus is difficult to extract as it is strongly bounded by the soil particles and its matrix.

Humus is an amorphous compounds with very high negatively charged organic sites such as those of carboxylic, phenolic, ketone, aldehyde groups. The negative charge on organic colloids is entirely dependent on the soil reaction which is low in highly acidic soils but exceeds its amount to that of silicate clays at pH >6. Humus saturated with H^+ on its exchange sites, can extract nutrient ions (e.g. Ca, Mg, K) from minerals by dissolving them, and then hold the nutrients on the exchange sites of the soil for plant uptake.

Chapter 19

Adsorption of Ions, Ion Exchange and its Significance

ADSORPTION OF IONS

The attraction or adhesion of ions or atoms or molecules of solid, liquid or gas to a surface of another molecule are called as adsorption of ions. It differs from absorption, as adsorption is a surface phenomenon whereas absorption takes place within the system. The sorption refers both adsorption and absorption while *desorption* is the reverse of adsorption i.e. detachment of adsorbed ions from system to the solution.

Ion exchange in soil: The exchange of ions between soil colloids and soil solution in contact is called ion exchange. It is a reversible process. If this exchange takes place between cations at adsorbed and solution phase of the soil, it is known as cation exchange but if this exchange takes place between two anions, it is known as anion exchange. Ion exchange in soils is important to plant nutrient availability and retention in soil.

The composition of soil solution is changing due to rainfall or irrigation, ion uptake by plants, leaching, fertilizer application, use of chemical amendments, sewage sludge and industrial effluents. This leads to exchange of ions from soil colloids to soil solution and vice versa

CATION EXCHANGE CAPACITY

"The capacity of soil to adsorb and exchange cations is called cation exchange capacity". The more the exchange capacity of cation for the soils, more will be the cations retained by the soil on its exchange complex.

The C.G.S. units of cation exchange capacity (CEC) of the soil is meq/100 g

The S.I. units of CEC is cmol(p^+)/kg soil

where (p^+) denotes proton.

Similarly, the exchange capacity of anions for the soil is given as meq/100 g or cmol(e^-) kg^1,

where (e^-) denotes electron.

The CEC of clay minerals ranges from 10-150 meq/100g soil, and organic matter from 200-400 meq/100g. Therefore, the CEC of the soils depends upon the amount of clay and its type and content of organic matter. Sandy soils have low CEC due small amount of clay and organic matter whereas clayey soils due to higher content of clay and organic matter have higher CEC.

The capacity of the soil to exchange for other cations varies with size of the cation and its hydrated radius, degree of hydration, valency, and concentration in soil solution.

The exchanging power of cations follows the order: $Na < K < NH_4 < Mg < Ca < H$

Importance of CEC

It plays an important role in soil fertility as exchange sites of the soils are saturated with basic cations such as Ca, Mg, K, Na and NH_4 ions. Thus, CEC is important in holding the cations on the exchange complex of soil and reducing their losses such as leaching from the soil. Amount of amendments for reclamation of acid (lime) and sodic (gypsum) soils depends upon CEC of the soils. Adsorption of heavy metals on exchange sites prevents them to pollute the groundwater.

Per cent base saturation

"It is defined as the per cent of total CEC occupied by the major base forming cations".

Factors affecting cation exchange equilibria in soils: The cation exchange equilibrium in soils is affected by both soil and solution factors:

Soil Factors:

i) **Clay minerals**: The specific surface area (SSA) and cation exchange capacity of clay minerals follow the order: montmorillonite > fine mica> kaolinite. But the specific charge density, SCD (ratio of CEC to SSA) of these clay minerals is reverse. The SCD is more important than CEC for describing the mono-divalent cation exchange equilibria.

ii) **Organic matter**: The presence of carboxyl and phenolic groups in organic compounds increases SCD of soil and favours the selectivity of multivalent cations. Organic matter also increases the ratio of internal: external surface area which favours the exchange of monovalent cations like K^+ and NH_4^+ influenced by the geometry of adsorption sites.

Solution factors

i) Concentration of soil solution: As the concentration of soil solution increases, the preference of soil for the divalent cations over the monovalent cations also increases.

ii) Type of cation: Cations of same valency but of different hydrated sizes also influence the preference of soils for the cations. As size increases, preference of the cation decreases.

iii) Type of anions: Presence of stronger undissociated ion pairs by divalent anions with divalent cations ($CaSO_4$) as compared to monovalent (Na_2SO_4) cations influence mon-divalent cation exchange equilibria in soils.

Anion exchange capacity: The anion exchange capacity of the soils varies with the clay minerals and its type, pH, and type of anions.

The anion exchange capacity of the clay minerals is reverse to that of CEC of clays i.e. kaolinite > montmorillonite > illite. It increases with the increase in acidity.

The affinity of adsorption of some anions in the soils follows the order: $PO_4^{-3} > SO_4^{-2} > Cl^- > NO_3^-$

Therefore, under normal pH, nitrate (NO_3), chloride (Cl^-) and sulphate (SO_4^{2-}) ions are easily leached down due to their less affinity to adsorption at exchange sites.

Importance of anion exchange

The phosphate anions due to its high affinity to adsorb or fix with the clay minerals reduces its availability in the soil and thus results in phosphorous fixation. Inspite of fixation of phosphate ions with clay minerals, it may also get fixed and form insoluble complexes with iron and aluminum oxides in acid soils and with calcium in calcareous or alkaline soils. The process is shown as:

$$\underset{\text{(soluble)}}{OH\text{–}clay + H_2PO_4^-} \rightleftharpoons \underset{\text{(insoluble)}}{H_2PO_4\text{–}clay + OH^-}$$

However, liming may reduce the phosphorous fixation by iron and aluminum oxide due to production of OH^- (hydroxyl ions) and increasing the pH of the acid soils.

At neutral pH value, the substantial amount of phosphorous still remains fixed with clay minerals. Under slightly acidic condition, the reverse reaction takes place and soluble phosphate is released, but under strongly acid soil, phosphate ions are fixed and become unavailable to plants.

Glossary

Abrasion: Wearing away due to rubbing against rough surface

Adhesion: Attraction between dissimilar particles

Adsorption: Attraction of substance on the surface

Aeolian: Materials transported by wind

Aerobic: Contains air

Alluvium: material carried by water

Amalgamation: mixing together of two or more things together

Amorphous: No definite form

Anaerobic: Absence of oxygen

Anhydrous: Without water

Apatite: Rock phosphate

Aspect: A part or feature of something

Bhangar: Old soils

Brittle: Hard but easily broken

Buffer: Resistance to change

Bulk density: Mass of bulk solid occupies a unit volume of a bed

Capillary rise: Movement of water upwards

Capillary: Narrow tube or voids

Catalyst: Substance that speed up the reaction

Chaotic: Disorganized

Cleavage: Mineral breakage along specific planes

Clod: A hard mass of soil formed artificially

Coarse textured: Bigger size particles

Coherent: Quality of holding together

Cohesion: Attraction between similar particles

Colloids: Very small particles (<0.001mm)

Colluvium: material transported by gravity

Compact: Hard layer

Conchoidal: A type of fracture resulting in smooth round surfaces

Coniferous: Cone bearing ever green tree

Crumb: A dry chunk of soil

Crust: A hard layer having high bulk density

Crust: Outermost solid shell of the earth

Crystal: Orderly arranged molecules

Crystalline: Ions arranged in an orderly manner

Deciduous: Trees that shed its leaves once in a year

Decomposition: The process of decaying

Dehydration: loss of water

Depression: Low lying area

Discrete: Separate or divided

Disintegration: Breakdown ion to smaller pieces

Diurnal: Daily fluctuations

Edaphology: Study of soil w.r.t. plants

Effervescence: Escape of gases from aqueous solution

Eluviation: Washing out of zone

Evapotranspiration: Loss of water from plants and soil

Exfoliation: Removal of surface layer

Fertility: Capacity of soil to supply nutrients

Field capacity: Water held in the soil after excess drainage

Flocculation: Joining together of the particles due to force of attraction

Foliation: Arrangement of minerals in layers

Friable: Pulverized

Gelatinous: Consistency of jelly

Genesis: Formation

Granular: Small spherical particles joined together to form porous material

Gritty: Coarse

Gypsum: Sulphate mineral found in deserts

Haphazard: Not orderly mannered

Heterogenous: Consisting of different substances

Humus: Dark coloured organic substance

Infiltration: Surface entry

Khadar: Young soil

Laminar: Travelling along smooth planes

Luster: Shining

Luster: Shining

Magma: Melted rocks

Malleable: Easily moldable

Massive: Very big

Matrix: Molecule or atoms arranged in rectangular array

Matrix: Set of numbers arranged in rows and columns to make a rectangular array

Metamorphism: Change of rocks in to some another rocks due to change in physical ad chemical form

Monophasic: Consisting of one phase

Morphology: form and structure

Particulate: Minute particles suspended in air

Penetration: To get in to it

Percolation: Movement from one layer to another

Permeable: Capacity to permits water

Petrology: Study of rocks

Polarize: To cause to divide

Pollutants: Material having undesirable effect

Polysaccharide: Long chain carbohydrate molecules

Porosity: Per cent of voids of the soil

Porous: Full of tiny holes or pores

Precipitation: Water released from clouds in the form of rain

Productivity: Capacity to produce

Pumice: Porous rock

Regolith: Material overlying the bed rock

Relief: Difference in elevation between the two points

Residue: Leftover after extraction

Resistance: Refusal to accept something

Runoff: Removal of water from surface

Sediment: Particles that settle at the bottom of the water

Sesquioxides: Atom and its oxide in the ratio of 2:3

Siderite: Mineral found in waterlogged soils

Solum: Soil

Specific gravity: Density of mineral w.r.t. density of water

Stratification: Arrangements of magma one over the other

Streak: A thin line or mark

Suction: The act of removing air from the space

Susceptible: Easily damaged

Susceptible: easily damaged by something

Suspension: Heterogeneous mixture of solid in liquids

Texture: A feel of something

Toposequence: Soils formed under different topography

Translocation: To move from one place to another

Translucent: Material that allow limited light to pass

Transparent: material through which light can pass

Trench: A long narrow hole dug in the ground for water to flow along

Vesicles: small tiny holes or pores

Virgin: unproductive

Viscosity: Resistance to flow

Viscous: Thick and sticky

Vitreous: Shine like of glass

Volatile: Highly unstable

Index